Albert Socin, Archibald Robert Stirling Kennedy

Arabic Grammar, Paradigms, Literature, Exercises and Glossary

Albert Socin, Archibald Robert Stirling Kennedy

Arabic Grammar, Paradigms, Literature, Exercises and Glossary

ISBN/EAN: 9783743361461

Manufactured in Europe, USA, Canada, Australia, Japa

Cover: Foto ©berggeist007 / pixelio.de

Manufactured and distributed by brebook publishing software (www.brebook.com)

Albert Socin, Archibald Robert Stirling Kennedy

Arabic Grammar, Paradigms, Literature, Exercises and Glossary

PORTA
LINGUARUM ORIENTALIU

INCHOAVIT

J. H. PETERMANN

CONTINUAVIT

HERM. L. STRACK.

ELEMENTA LINGUARUM

Hebraicae, Phoeniciae, Biblico-Aramaicae,
Samaritanae, Targumicae, Syriacae, Arabicae
Aethiopicae, Assyriacae, Aegyptiacae, Copticae
Armeniacae, Persicae, Turcicae, aliarum

studiis academicis accommodaverunt

J. H. *Petermann*, H. L. *Strack*, E. *Nestle*, A. *Socin*, F. *Praet*,
A. *Merx*, *Aug. Mueller*, *Friedr. Delitzsch*, C. *Salemann*
Ad. Erman, V. *Shukovski*, *Th. Noeldeke*,
G. *Steindorff*, R. *Bruennow*, *Dav. H. Mueller*, G. *Jacob*

PARS IV.

ARABIC GRAMMAR

BY

A. SOCIN.

SECOND EDITION.

BERLIN,
REUTHER & REICHARD

LONDON
WILLIAMS & NORGATE
14, HENRIETTA STREET

NEW YORK
B. WESTERMANN & C
812, BROADWAY.

1895.

ARABIC GRAMMAR

paradigms, literature, exercises
and
glossary

by
Dr. A. SOCIN

Second English edition
Translated from the third German edition
by the
Rev. Arch. R.S. KENNEDY D.D.

Berlin
Reuther & Reichard
1895

PREFACE

TO THE

SECOND ENGLISH EDITION.

The aim of the following pages is to furnish intending students of classical Arabic with the most important rules both of the Accidence and of the Syntax in the briefest possible form. The present edition, the second in English, is a translation of the third German edition of 1894, to which, save for a few corrections and additions, it in all respects corresponds. Its German counterpart has been considerably altered compared with the second edition because of the publishers' intention to issue a separate chrestomathy of Arabic prose. Professor R. Bruennow, a scholar of approved ability, was entrusted with the preparation of this work which appeared in the year 1894. The connected narratives which formerly composed the chrestomathy of the grammar were, according to arrangement, incorporated in Bruennow's work, and consequently had to be dropped from the new edition of the grammar. On the other hand the latter was now extended, more particularly in the part dealing with the syntax, with the result that it will now be found, with few exceptions, to be sufficient for the understanding of the new chrestomathy. At the same time, the fact must again be emphasised that the present work does not pretend to take the place of any of the larger treatises; the English student who wishes to advance beyond the elements of Arabic must have recourse to the latest edition (the third) of Wright's excellent grammar. For this reason the author has deemed it his duty to adhere to his former view and to decline, in a book intended for beginners, to enter into the technical terminology of the Arab grammarians—which may safely be left to the larger grammars; still the Arabic specialist will easily discover that their views have been taken into account even in the present elementary work. The best introduction to this department of study will be found to be the reading of the Ağrumīye, which Bruennow has printed in his Chrestomathy.

In order to lighten the first lessons in grammar, the exercises consisting of short sentences and anecdotes have been increased by the addition of a few short stories, by means of which a sort of stepping stone is provided to the prose chrestomathy.

The passages for translation into Arabic have been retained unchanged along with the appropriate glossary. Experience has shown that this part of the chrestomathy has unquestionably been of service; and although I am strongly of opinion that this class of exercises is of real value in such systematic instruction as is necessary at first, I am in no wise blind to the difficulties which the correction of such exercises entails even on the teacher of Arabic. In order to meet such difficulties, I have selected single sentences and anecdotes from Arabic authors, and have so arranged both notes and glossary that the student, who in any case will have to make diligent use of grammar and dictionary, is so to say compelled to reproduce exactly the Arabic original. From what has just been said, it is clear that this part of the book, at least, presupposes a teacher, for I am convinced that the grammar of Arabic as a whole, and the syntax in particular, can only be mastered with extreme difficulty by self-instruction. I would add, however, that translation from English should be taken at first in the smallest possible doses, and even in this way only after the student has read a part of the Arabic texts.

The synopsis of Arabic literature has also been extended. Strictly speaking, this section is out of place in an elementary work; still it may afford a stimulus to a beginner here and there, and supply an occasional hint to those pursuing the study of Arabic by themselves, or at a distance from the larger seats of learning.

The present English edition is an entirely new translation. This difficult and tedious work has been undertaken by Professor Archd. R. S. KENNEDY of Edinburgh University. To him and to his late assistant, Mr. W. B. STEVENSON B. D., Vans Dunlop Scholar in Semitic Languages of the same University, who has rendered us great assistance in the reading of the proofs, I cannot omit to express here my warmest thanks for their co-operation.

A. SOCIN.

NOTE BY THE TRANSLATOR.

I have only to add to the foregoing, that my responsibility as translator does not extend to the English-Arabic exercises and the relative glossary. A few verbal changes excepted—chiefly where the "violence done to the Queen's English" (p. 57*) was greater than seemed absolutely necessary—these have been reprinted from the first edition. I have also inserted an additional reference here and there, and in the bibliographical section I am responsible for one or two additional entries.

I wish also to express my personal indebtedness to Mr. Stevenson, without whose generous co-operation, owing to my absence in the East, the book would not have been ready in time for this winter's work.

20th September 1895.

A. R. S. K.

TABLE OF CONTENTS.

GRAMMAR.

I. ORTHOGRAPHY AND PHONOLOGY (§§ 1—11).

		Page
§ 1.	Consonants	1
§ 2.	Long Vowels	6
§ 3.	Short Vowels, Nunation, Gezma	8
§ 4.	Hamza	9
§ 5.	Tešdīd	11
§ 6.	Waṣla	12
§ 7.	Medda	15
§ 8.	The Syllable	16
§ 9.	The Tone	17
§ 10.	Pause	18
§ 11.	The Arabic Cyphers and Contractions	18

II. ACCIDENCE (§§ 12—96).
Chap. I. The Pronoun (§§ 12—15).

§ 12.	Personal Pronouns	19
§ 13.	Demonstrative Pronouns	21
§ 14.	Relative Pronouns	23
§ 15.	Interrogative Pronouns	24

Chap. II. The Verb (§§ 16—54).

§ 16.	Groundform	24
§ 17.	Conspectus of the derived Stems	24
§ 18.	I. Stem	25
§ 19.	II. Stem	26
§ 20.	III. Stem	26
§ 21.	IV. Stem	27

		Page
§ 22. V. Stem		27
§ 23. VI. Stem		28
§ 24. VII. Stem		28
§ 25. VIII. Stem		28
§ 26. IX. and XI. Stems		29
§ 27. X. Stem		29
§ 28. The Stems of the quadriliteral Verb		30
§ 29. The Passive		30
§ 30. The Tenses		30
§ 31. The Moods		31
§ 32. Imperative		32
§ 33. Inflexion for Person and Number		33
§§ 34—36. Verbs mediae geminatae		34
§§ 37—38. Verba hamzata		36
§ 39. Weak Verbs		38
§ 40. Verba primae و et ى		38
§§ 41—44. Verba mediae و et ى		39
§§ 45—48. Verba ultimae و et ى		41
§ 49. Doubly weak Verbs		44
§ 50. The Verb لَيْسَ		45
§ 51. Verbs of Praise and Blame		46
§ 52. Forms of Admiration		46
§ 53. The Verb with Pronominal Suffixes		46
§ 54. Sign of the Accusative		47

Chap. III. The Noun (§§ 55—90).

a. Formation of Nouns.

§ 55. Primitive and derived Nouns		48
§ 56. Summary of the simple Nouns		49
§ 57. Nouns with Preformatives		49
§ 58. Nouns with Afformatives		50
§ 59. Quadriliteral Nouns		50
§ 60. Participles		50
§ 61. Infinitives		51
§ 62. Verbal Adjectives		53

XII   Contents.

	Page
§ 63. Intensive Forms	54
§ 64. Nomina loci, instrumenti, speciei	55
§ 65. Nomina relativa	56
§ 66. Nomina deminutiva	57
§ 67. Nouns from Stems mediae geminatae	57
§ 68. Nouns from Stems with Hamza	58
§ 69. Nouns from Stems primae و	58
§ 70. Nouns from Stems med. و and ى	59
§ 71. Nouns from Stems ultimae و and ى	60

b. Gender of Nouns.

§ 72. Masculine and Feminine Gender	62
§§ 73—74. Formation of the Feminine	63

c. Inflexion of Nouns.

§ 75. Number and Case	65
§ 76. Formation of the Dual and Plural	66
§ 77. Case-endings of Singular. Triptote and Diptote Nouns	67
§ 78. Diptotes	68
§ 79. Inflection of the Determined Noun	68
§ 80. Shortening of Dual and Plural in the Construct State	69
§ 81. Inflection of Nouns in *in* and *an* from ult. و and ى	70
§ 82. The Noun with the Pronominal Suffixes	71
§ 83. Vowel Changes in the Pluralis Sanus	72
§ 84. Proper Names compounded with ابن	73
§ 85. Vocative	73
§ 86. Collective Nouns	74
§ 87. Broken Plurals	75
§ 88. List of the principal varieties of the Broken Plural	76
§ 89. Broken Plurals from Quadriliteral Nouns	78
§ 90. Nouns of irregular Formation	80

Chap. IV. *The Numerals* (§§ 91—93).

§ 91. The Cardinal Numbers	83
§ 92. The Connection of the numeral with the thing numbered	85
§ 93. Ordinal Numbers and Fractions	86

CONTENTS. **XIII**
Page

Chap. V. *Particles* (§§ 94—96).

§ 94. Adverbs, Prepositions, Conjunctions 88
§ 95. Inseparable Particles 88
§ 96. Prepositions and Particles with Suffixes 89

III. SYNTAX (§§ 97—160).

Chap. I. *Tenses and Moods* (§§ 97—104).

§ 97. Perfect and Imperfect 90
§ 98. Use of the Perfect 91
§ 99. Use of the Imperfect 92
§ 100. Subjunctive 94
§ 101. Modus apocopatus 95
§ 102. Modus energicus 95
§ 103. Passive 96
§ 104. Participles 96

Chap. II. *Government of the Verb* (§§ 105—117).

§ 105. The Verb and its Compliment 97
§ 106. Accusative 97
§ 107. Accusative after verbs of coming &c. 97
§ 108. Verbs with two Accusatives 97
§ 109. The Absolute Object 98
§ 110. The Accusative as Predicate 99
§ 111. Accusative with ي 100
§ 112. Accusative with و of Concomitance 101
§ 113. Accusative of nearer Definition 101
§ 114* Accusative in Exclamations 102
§§ 114—116. The Verb with Prepositions 103
§ 117. ل in Dates 104

Chap. III. *Government of the Noun* (§§ 118—134).

§ 118. The Noun with the Article (Determination) . . 105
§ 119. Apposition 106
§§ 120—122. Qualifying Adjuncts 107
§§ 123—130. The Genitive Relation 109

		Page
§ 131. The Construction of the Infinitive	. . .	112
§ 132. The Participle and its Object . .	. . .	113
§ 133. The Nomen Regens undetermined	. . .	114
§ 134. Improper Annexation		114

Chap. IV. *The Simple Sentence* (§§ 135—151).

§ 135. Distinction bet. Nominal and Verbal Sentences .	115
§ 136. The Verb in the Verbal Sentence	115
§§ 137—138. Indefinite Subject	117
§ 139. The Predicate in the Nominal Sentence	118
§§ 140—146. Connection bet. Subject and Predicate . .	119
§ 147. The Particles *'inna* and *'anna*	122
§ 148. Subordinate Sentences	123
§ 149. More than one Predicate	125
§ 150. Negative Sentences	125
§ 151. The Particle of Exception	126

Chap. V. *Compound Sentences* (§§ 152—161).

§ 152. Co-ordinate Sentences	127
§§ 153—156. The Relative Clause	128
§ 157. The Circumstantial Clause	131
§ 158. The Temporal Clause	132
§ 159. The Conditional Clause with the Perfect . . .	133
§ 160. The Conditional Clause with the Apoc. Impf. .	134
§ 161. The Particle ف in the Apodosis	134

APPENDIX.

Computation of Time (Names of the Days of the Week, the Months &c.) 136

LITERATURE.

A. Bibliography	139
B. Introduction and general	144
C. Chrestomathies	144
D. Grammars	145

	Page
E Lexicography	147
F. Koran, Islam, Life of Muhammed, Bible &c.	150
G. Jurisprudence	153
H. Philosophy	154
I. Natural Sciences and Medicine	156
K. History, Biographies	157
L. Cosmography, Geography, Ethnography, Travels	163
M. Poetry	166
N. Belles Lettres, Ethics, Romances	169

PARADIGMS.

I. Suffixes and Prefixes for the Conjugation of the Verb	3*
II. Strong triliteral Verb Act. I.	4*
III. Strong triliteral Verb Pass. I.	6*
IV. Quadriliteral Verb, derived Stems	7*
V. Strong triliteral Verb, derived Stems	8*
VI. Verbum mediae geminatae Act. I.	10*
VII. Verbum mediae geminatae Pass. I.	11*
VIII. Verbum mediae geminatae, derived Stems	12*
IX. Verba hamzata	13*
X. Verbum primae radicalis و et ى	14*
XI. Verbum mediae radicalis و Act. I.	15*
XII. Verbum mediae radicalis ى Act. I.	16*
XIII. Verbum mediae radicalis و vel ى Pass.	17*
XIV. Verbum med. radicalis و et ى, derived Stems	18*
XV. Verbum tertiae radicalis و (فَعَلَ) Act. I.	19*
XVI. Verbum tertiae radicalis ى (فَعَلَ) Act. I.	20*
XVII. Verbum tertiae radicalis و vel ى (فَعَلَ) Act. I.	21*
XVIII. Verbum tertiae radicalis و vel ى Pass. I.	24*
XIX. Verbum tertiae radicalis و vel ى, derived Stems	22*
XX. Nomen generis masculini	25*
XXI. Nomen generis feminini	26*
XXII. Nouns in "in" and "an"	27*
XXIII. The Noun with Pronominal Suffixes	28*

EXERCISES AND TEXTS.

		Page
I. Exercises on the Grammar		
A. For practice in Reading	. . . .	30*
B. Exercises on the Accidence	. . . . . .	32*
C. Exercises on the Syntax	. . . . . .	35*
II. Connected Extracts	. . . .	48*
III. For Translation into Arabic	. . . .	57*

GLOSSARIES.

A. English-Arabic	. . . .	79*
B. Arabic-English	. . . .	104*

PART I.

GRAMMAR

AND

BIBLIOGRAPHY.

Socin, Arabic Grammar.[2]

GRAMMAR.

I. THE ARABIC CHARACTERS. PHONOLOGY
(§§ 1—11).

1.
a.
The Consonants. The Arabs at first used the Syriac characters and the Syriac alphabet, in which the order of the characters is the same as in Hebrew. A relic of this earlier order is still preserved, in the employment—afterwards seldom resorted to—of the letters of the alphabet as cyphers (cf. pages 4—5). At an early period, however, the Arabs distinguished by means of diacritical points a number of sounds which were not so distinguished in the older alphabet. By a process of curtailment, moreover, a number of characters became so like each other that they had to be distinguished by similar diacritical signs. The next step was to group together in the alphabet the characters which in this way had come to resemble each other. Hence the Arabic alphabet now consists of twenty-eight consonantal signs, the usual order and

TABLE OF CHARACTERS.

Names	Form				Value			Hebrew
	Not joined.	Joined only to preceding cons.	Join to precedg. and follg. cons.	Joined only to following cons.	Pronunciation	Transcription	Numerical Value.	
1 اَلِفْ Alif*	ا	ـا	—	—	cf. §§ 2 and 4		1	א
2 بَاءْ Bā	ب	ـب	ـبـ	بـ	b	b	2	ב
3 تَاءْ Tā	ت	ـت	ـتـ	تـ	t	t	400	ת
4 ثَاءْ Thā	ث	ـث	ـثـ	ثـ	English hard th as in *thing*	ṯ	500	
5 جِيمْ Jīm	ج	ـج	ـجـ	جـ	orig. g hard; later g in Italian *giorno*; English j	ǧ	3	ג
6 حَاءْ Ḥhā	ح	ـح	ـحـ	حـ	strong h with friction of larynx as if wheezing	ḥ	8	ח
7 خَاءْ Khā	خ	ـخ	ـخـ	خـ	ch in Scotch *loch*	ḫ	600	
8 دَالْ Dāl	د	ـد	—	—	d	d	4	
9 ذَالْ Dhāl	ذ	ـذ	—	—	soft th, as in *this*	ḏ	700	ד
10 رَاءْ Rā	ر	ـر	—	—	r	r	200	ר
11 زَايْ Zāi	ز	ـز	—	—	z as in *zeal*; soft s as in *rose*	z	7	ז
12 سِينْ Sīn	س	ـس	ـسـ	سـ	hard s	s	60	ס
13 شِينْ Šīn	ش	ـش	ـشـ	شـ	sh	š	300	שׁ

* The termination ٌ i. e. *un* (see § 3 b) is neglected in the transliteration, as in the modern Arabic pronunciation.

TABLE OF CHARACTERS.

Names	Form				Value			
	Not joined.	Joined only to preceding cons.	Join to precedg. and follg. cons.	Joined only to following cons.	Pronunciation	Transcription.	Numerical Value.	Hebrew.
14 صَاد Ṣād	ص	ـص	ـصـ	صـ	emphatic s	ṣ	90	צ
15 ضَاد Ḍād	ض	ـض	ـضـ	ضـ	emphatic d (tongue pressed against the gum)	ḍ	800	
16 طَا Ṭā	ط	ـط	ـطـ	طـ	emphatic t	ṭ	9	ט
17 ظَا Ẓā	ظ	ـظ	ـظـ	ظـ	emphatic z	ẓ	900	
18 عَيْن ʿAin	ع	ـع	ـعـ	عـ	produced by a tightening of the violently compressed glottis	ʿ	70	ע
19 غَيْن Ghain	غ	ـغ	ـغـ	غـ	guttural r	ġ	1000	
20 فَا Fā	ف	ـف	ـفـ	فـ	f	f	80	פ
21 قَاف Ḳāf	ق	ـق	ـقـ	قـ	deep emphatic k	ḳ	100	ק
22 كَاف Kāf	ك	ـك	ـكـ	كـ	k	k	20	כ
23 لَام Lām	ل	ـل	ـلـ	لـ	l	l	30	ל
24 مِيم Mīm	م	ـم	ـمـ	مـ	m	m	40	מ
25 نُون Nūn	ن	ـن	ـنـ	نـ	n	n	50	נ
26 هَا Hā	ه	ـه	ـهـ	هـ	h	h	5	ה
27 وَاو Wāw	و	ـو	—	—	w	w	6	ו
28 يَا Yā	ي	ـي	ـيـ	يـ	y	y	10	י

forms of which are exhibited on pp. 4—5. These signs are written and read from right to left. Some are joined, to the letters preceding or following, others are not, as indicated in the table referred to.

b. When ة (No. 26), at the end of a word, indicates the feminine termination (§ 73), two dots are placed over it to show that it must be pronounced as *t* (Nr. 3); thus: ة.

c. Very frequently, especially at the beginning of words, certain letters, instead of being written alongside of each other, are placed one above the other this is particularly the case with the letters ح ج خ (Nos. 5—7), e. g. جـ for جب (Nos. 2 and 6), حج (Nos. 5 and 6) for حج, جح (Nos. 18 and 5) for عج &c. Instead of لا (Nos. 23 and 1) the Arabs write لا or لا (the Lām in the latter form beginning at the left of Alif).

2. *The vowels, how indicated.* In the earliest times
a. the Arabs indicated only the long vowels *ā*, *ī*, *ū*, and the diphthongs *au*, *ai* (whose second element they regarded as a consonant); this was done by employing the sign ا (No. 1) for *ā*, و (No. 27) for *ū* and (with *a*) *au*, ى (No. 28) for *ī* and (with *a*) *ai*. In cases where و and ى indicate the sounds *au* and *ai*, which we pronounce as diphthongs, Sukūn (see § 3 *c*) is ordinarily placed

over these letters, to denote that they have no vowel of their own. Examples: قَال ḳāla, سِير sīra, سوق sūḳun, بَيْع baiʿun, نَوْم naumun.

In the oldest writing, the long *ā* was not uniformly represented by ا, but was occasionally left unrepresented. This omission has continued to be observed in a series of very common words; in such cases, however, an upright stroke is usually placed over the consonant that is to be pronounced with the long *ā*, e. g. هٰذَا (for هاذا) hāḏā, الٰه 'ilāhun (God), رحمٰن raḥmānu. Frequently, however, in our printed editions, we find this long *ā* represented by a simple ـٰ , thus: هَذَا hāḏā. *b.*

In a few words a ‌و after an *a* does not indicate the pronounciation *au* but a long *ā*, originally no doubt an obscure *ā̊*; in this case, too, the upright stroke is the usual sign, e. g. حيوٰة (alongside of حياة) ḥayātun life (but ا with Suffixes: حياته ḥayātuhu his life). *c.*

At the end of many words ى is likewise employed to represent a long *ā*; in such cases (like the ‌و in *c*) it does not receive the Sukūn (§ 3 *c*), e. g. رمى (or رمىٰ) ramā (he has thrown); in the middle of a word, on the other hand, ا takes the place of this ى; thus with a suffix رماه ramāhu he has thrown it. *d.*

3. THE SHORT VOWELS.

NOTE a. In a few rare cases, in the middle of a word, we find $\bar{a}$ denoted by ىٰ, as in the foreign word تَوْرِيَةٌ *taurātun* Torah.

NOTE b. Should ىٰ be preceded by a ي, ا is written for the former in order to prevent two ي coming together; e. g. دنيا *dunyā* world für دنيى (§ 74 a).

e. Occasionally an ا is added to a final $\bar{u}$ or *au*, but it is entirely left out of account in the pronunciation; e. g. كتبوا *katabū*, رموا *ramau* (§§ 33 and 53).

3. *The short vowels* were originally, as a rule, left
a. unrepresented[1]; afterwards the following signs were employed to represent the short vowels, and (in conjunction with the signs discussed in § 2) the long vowels as well:

1) فَتْحَة Fatḥa[2] (also فَتْح Fatḥ) ◌َ for *a* (in certain cases to be pronounced like *e* in men, also like German *ä* in Männer), e. g. قَتَلَ *katala*, قَالَ *kāla*.

2) كَسْرَة Kesra (also كَسْر Kesr) ◌ِ for *i*, e. g. غَضِبَ *ġaḍiba*, يَبِيعُ *yabīʿu*.

3) ضَمَّة Ḍamma (also ضَمّ Ḍamm) ◌ُ for *u*, e. g. يَكْتُبُ *yaktubu*; يَفُوتُ *yafūtu*.

b. When these signs for the short vowels are written

---

[1] Many books, particularly those printed in the East, are printed without these vowel signs.

[2] The terminations ةً , ةٌ are here represented in the transliteration by *a*, as in modern Arabic.

twice at the end of a word, they are to be pronounced with a final *n* (called by the Arabs تَنْوِين Tanwīn, by us frequently Nunation, from the letter nūn), e. g. شَمْسٍ *šamsin*, رَجُلٌ *raǧulun*. The Nunation *an* receives as an additional indication the letter ا, but the pronunciation remains unaffected, e. g. مَالًا *mālan*. This ا is omitted only when the Nunation accompanies the feminine termination ة (see above § 1 *b*), e. g. مَرْكُوبَةً *markūbatan*, or in cases where the word already has a final ا, or in its place a ى quiescing in *a*, e. g. رِبًا *riban*, هُدًى *hudan* (§ 2 *d*). The same holds good in most cases after Hamza (§ 4). The vowel of the Nunation is always short.

When a consonant has no vowel of its own, this is indicated by the sign ـْ جَزْمَة Ǧezma (also called سُكُون Sukūn [Rest]) e. g. سَافَرْتُ *sāfartu*, مَشَيْتُ *mašaita* (cf. § 2). On the omission of this sign see § 5. A consonant which is thus pronounced without a vowel following is said to be "resting".

Hamza. In order to distinguish the cases in which ا was employed to denote *ā* (§ 2), from those in which it had (as originally in Hebrew) its proper force as a consonant, the Arabs gave it the additional sign

4. *Hamza.*
a.

4. HAMZA.

هَمْزَة Hamza ء (in form a modified ع ʿAin). ﺍ accordingly denotes the closure of the larynx by which the breath, engaged in voice production, is turned on or off, according as the Hamza precedes or follows a vowel. It is best heard in English before the second of such pairs of words as "sea eagle", "mine eyes". Its effect may also be noted by comparing the two following pronunciations of Ḳurān, viz: Ḳu-rān and Ḳur-'ān (the latter with Hamza). In the transliteration we indicate ﺍ by ' except at the beginning of a word where no indication is required. The sign ء is placed under the ﺍ, when followed by an *i*-sound. Examples أَمْرٌ *'amrun*, إِبِلٌ *'ibilun*, أُمَمٌ *'umamun*; سَأَلَ *sa'ala*, رَأْسٌ *ra'-sun*; اِقْرَأْ *iḳ-ra'*. In the last two examples Hamza closes the syllable.

b.   Before or after an *i*- or *u*-sound, the signs و and ى are generally employed instead of ﺍ as the bearers of the Hamza, in which case ى is written without the two dots: e. g. بَؤُسَ *ba'usa*, يُؤْثَرُ *yu'ṯaru*, يُؤَاثَرُ *yu'āṯaru*: جِئْتَ *ji'ta*, صَئِبَ *ṣa'iba*, يُبَارِئُ *yubāri'u*.

c.   After a long vowel, and in most cases after Sukūn, Hamza as a rule has no bearer, but is written on or above the line, thus: إِرْضَاءٌ *'irḍā'un*, بَرْءٌ (or بَرْأٌ) *bar'un*. After a vowelless consonant in the middle of a word, Hamza with its vowel is placed over the connecting stroke (except, of course, when no such connection is possible, as after و, in مَمْلُوءَةٌ *mamlū'atun*), thus: خَطِيْئَةٌ *ḥaṭī'atun*; in the same way شَيْئًا *šai'an*; On the other hand, in cases like إِمْضَآءً *'imḍā'an* no Alif is written at the end, cf. § 3 *b*.

Tešdīd. That a consonant is to be sounded twice is indicated by the sign of doubling ّ, named تَشْدِيدٌ Tešdīd or شَدٌّ Šedd (from the initial ش of this word the sign ّ has been derived), e. g. سَبَّ *sabba*, تَرَحُّلٌ *taraḥḥulun*. This doubling of a consonant is either due —as in the examples just given—to the essential nature of the form, nominal or verbal (as for example the verbal forms corresponding to the Hebrew *Pi"el* § 19), or is the result of assimilation.

When one consonant is assimilated to another, the assimilation is further graphically represented by the removal of the Sukūn from the assimilated consonant. This applies to the *l* of the article أَل, when the latter precedes one of the following consonants: ت, ث, د, ذ, ر, ز, س, ش, ص, ض, ط, ظ, ل, ن (that is, dentals, sibilants and *r, l, n*). Examples: اَلتَّاجِرُ *attāǧiru*, اَلتَّلْجُ *attalǧu*, اَلشَّمْسُ *aššamsu*, (the sun), but اَلْقَمَرُ *alḳamaru* (the moon). From the fact that the two last examples are stereotyped those consonants that may be assimilated are technically called *solar* letters, those that do not admit of assimilation. *lunar* letters.

NOTE a. The word اِلٰه 'ilāhun, God, when joined to the article drops the first syllable and becomes اَللّٰهُ (§ 2 b) *allāhu*.

NOTE b. The words مِنْ *min*, عَنْ *'an*, اَنْ *'an* (and اِنْ *'in*), when followed by a few words beginning with م *m* or ل *l* are usually combined with them into one word, the final ن *n* being at the same time assimilated to the following consonant, e. g. مِمَّ *mimmā* from مِنْ مَا *min mā*, اَلَّا *'allā* from اَنْ لَا *'an lā*.

6. ***Waṣla*** ـٰ. A word beginning with two consonants
a. receives in Arabic either a full helping-vowel preceded by Hamza in accordance with § 4 *a* (e. g. أَفْلَاطُون Plato), or merely a vowel which is heard only when the word is standing *alone*, but which must be given up when the word in question comes to stand after another word in the sentence. Thus in the latter case we find أُقْتُلْ *uḳtul* instead of قْتُلْ *ḳtul*. The ا which is prefixed in this and similar cases is, however, still written although the helping vowel accompanying it is given up, and it then receives over it the sign وَصْلَة *waṣla* ـٰ e. g. بِنْتُ الْوَزِيرِ *bintulwazīri*. The two words thus united together are also to be pronounced as if they formed a single word. Such an *Alif Waṣlatum* or Waṣla-bearing Alif is called a *connective* Alif in

contradistinction to a *disjunctive* Alif, that is, an *Alif hamzatum* or Hamza-bearing Alif (cf. § 4).

NOTE. The sign ࣱ is a modification of ص; waṣla or ṣila denotes "close connection".

When a connective Alif has to be employed at the beginning of a sentence, a full vowel must be pronounced, but, as written, only the proper vowel sign may accompany the Alif, never a Hamza. Thus we have اَلرَّسُولُ *arrasūlu*, اُخْرُجْ *uḫruǧ* but قَالَ اخْرُجْ pronounce *ḳālaḫruǧ*.

In the last example the division of the syllables is now *ḳā-laḫ-ruǧ*. If the vowel preceding a connective Alif is long, it must now be pronounced as a short vowel, since it stands in a shut syllable (see § 8). Thus فِي ٱلْفُلْكِ, properly *fī-lfulki*, has now the following syllables *fil-ful-ki*; so too رَضِيَ ٱللّٰهِ *riḍa-llāhi* (§ 2 d) = *ri-ḍal-lā-hi*, ذَبَحُوا ٱلْإِوَزَّ (§ 2 e) *ḍa-ba-ḥul-'iwazza*.

If the word before a connective Alif ends in a consonant which has no vowel of its own, the consonant receives a helping-vowel. The most natural vowel in such a case is *i*, e. g. ضَرَبَتِ ٱلْعَبْدَ *ḍarabati-l'abda* (for ضَرَبَتْ); so اِسْتِقْبَالٌ *istiḳbālun* with the article اَل: اَلِٱسْتِقْبَالُ *alistiḳbālu*, in syllables thus: *a-lis-tiḳ-bā-lu*. In certain cases original final vowels that have been dropped reappear before the connective Alif, e. g. هُمْ ٱلْكَافِرُونَ *hu-mul-kā-fi-rū-na*. The first word is otherwise uniformly هُمْ *hum* (§ 12 *a*). — The Nunation (§ 3 *b*) is also treated as if it ended in a consonant; the favourite vowel in this case is *i*, e. g. رَجُلٍ ٱسْمُهُ pronounced as if written رَجُلُنِ ٱسْمُهُ *raǧulunismuhu*, in syllables: *ra-ǧu-lu-nis-mu-hu*.

> NOTE. Before a connective alif the preposition عَنْ "away from" becomes عَنِ, the preposition مِنْ, "from" becomes مِنَ, but before the article مِنْ.

e. The same rule applies to a word ending in a so-called diphthong (cf. § 2); the consonant (و or ي) forming the second part of the diphthong must receive a helping vowel before a connective Alif, which vowel is *u* or *i* according as the consonant in question is و or ي. Thus we have مُصْطَفَوُ ٱللّٰهِ *muṣ-ṭa-fa-wul-lā-hi* in place of مُصْطَفَوْ ٱللّٰهِ, رِجْلَيِ ٱلْبَقَرَةِ *riǧ-la-yil-ba-ḳa-ra-ti* for رِجْلَيْ ٱلْبَقَرَةِ. (So, too, with the termination ـَوْا § 2 *e*).

> NOTE. The particles أَوْ "or" and أَوْ "would that!" take *i* as helping vowel.

f. The connective Alif is altogether omitted in the following cases:

1) In the article اَلْ, when it receives as prefixes the particles لِ *li* or لَ *la*; e. g. لِلْحَقِّ, *lil-ḥaḳ-ḳi* for لِاَلْحَقِّ, لَلْمَجْدُ *lal-maǧ-du* for لَاَلْمَجْدُ.

2) In اِبْنُ son, in apposition to the proper name of the son and followed in the genitive by the name of the father; e. g. مُسْلِمُ بْنُ ٱلْوَلِيدِ *mus-li-mub-nul-wa-lī-di* Muslim, the son of al-Walīd. At the beginning of a line, however, even in this case we must write اِبْنُ.

3) In the word اِسْمٌ *ismun*, name, after the preposition بِ *bi* in the oft recurring formula بِسْمِ ٱللّٰهِ *bis-mil-lā-hi*, in the name of God.

7. *Medda.* Inasmuch as the Arabic orthography cannot tolerate two Alifs side by side, in such a case only a single Alif is written, over which is placed a مَدَّةٌ Medda or Medd ~ (a sign derived from مَدّ). At the beginning of a word or syllable the Medda carries with it the force of a Hamza; the vowel sign Fatḥ is then also dropped, e. g. آكِلٌ *'ā-ki-lun* for أَأْكِلٌ, قُرْآنٌ *ḳur-'ā-nun* for قُرْأَانٌ; so آمَنَ *'ā-ma-na* for أَأْمَنَ, since the Hamza of the second Alif disappears as explained § 38 *a*.

NOTE. رَاىٰ *ra'ā* may be taken as an example of the rule just given. With suffixes it ought to appear, according to § 2d, as رَاىُ, which, however, is written رَآهُ in syllables *ra-'ā-hū*.

b. Since a ء after a long *ā* ١ـ is written on the line (§ 4c) without receiving an Alif as bearer, the ا preceding the Hamza in such cases likewise receives Medda, as a rule, although the latter has no effect on the pronunciation of the word, e. g. جَآءَ *ğā-'a* (for جَاءَ), تَفَآءَلُوا *ta-fā-'a-lū*; and the same where و or ى appears as the bearer of Hamza أَحِبَّآؤُ *a-ḥib-bā-'u-hu*, قَآئِلٌ *ḳā-'i-lun*.

NOTE. Arabic orthography has also an objection to two Waws appearing side by side, if the first has a Ḍamma (even though the first may be only the bearer of a Hamza, as expained in § 4b). Thus رُؤُوسٌ *ru'ūsun* is often written رُؤُسٌ.

8. *The Syllable.* An *open* syllable ends in a vowel short or long; a *shut* syllable ends in a consonant. Every syllable begins with a single consonant, not with two or more (cf. § 6). A *short* syllable consists of a consonant with a short vowel, as in the second syllable of مَاتَ *mā-tă* (with two open syllables); a *long* syllable consists either ① of a consonant with a long vowel, like the open syllable *mā* in the above example, or ② of a consonant, a short vowel and a consonant (shut syllable) e. g. both the syllables of

قَتْلٌ *ḳat-lun* (so too مَوْتٌ *mau-tun* § 2 *a*) شَرٌّ *šar-ran*, or (3) of <u>a shut syllable with a long vowel</u>. This last variety, however, is only found (exclusive of pausal effects § 10) when the following consonant has been doubled (§ 5) and is preceded by a long *ā*, as in دَابَّةٌ *dāb-ba-tun* (rarely after *ai* as in ذُوَيْبَّةٌ *du-waib-ba-tun* which is derived according to § 66 from *dābbatun*). Such a syllable may be described as doubly long. Other syllables of this sort are shortened as يَقُلْ *yaḳul* from يَقُولُ *yaḳūl*; رَمَتْ *ramat* from رَمَاتْ *ramāt*.

NOTE. A word consisting of but one short syllable, if it stands alone, either receives an addition at the end (see § 49 *ab*), or is joined to the following word. The latter method is adopted in a series of particles (see § 94), which notwithstanding the connection are still regarded as more or less independent words. The principal stress, however, rests on the words with which the particles are connected.

9. *The Accent or Tone.* The accent in Arabic is thrown backwards towards the beginning of the word till it meets a *long* syllable, or if there is no such syllable, till it reaches the first syllable of the word. A simple long syllable at the end of a word, however, <u>does not receive the accent</u>. Examples of words with a final short syllable: ضَارَبَ *ḍáraba*, اِسْتَنْكَرَ *istánkara*; with a final long syllable: تَمَامْتُمَا *tamámtumā*, فَرْدٌ *fárdun*, مَمْلَكَةٌ *mámlakatun*, ضَرَبُوا *ḍárabū*, لِدَةٌ *lídatun*.

18 10. PAUSE. 11. NUMERICAL SIGNS. ABBREVIATIONS.

Exceptions: A syllable with a connective Alif (§ 6), as in اُقْتُلْ (see § 6 a), cannot receive the accent; the pronunciation is therefore *uḳtŭl*. In the same way monosyllabic inseparable particles, like وَ and فَ (cf. § 94), prefixed to words, do not affect the accentuation of the latter; e. g. فَمَشَى *famáša*.

10. In *pause* final short vowels are dropped. Also
(1) the Nunation *un* and *in*; the Nunation *an* is changed
(2) to *ā*, the feminine termination ةً to هْ (with the *h*
(3) sounded): thus نَازِلُونْ *nāzilūn* for نَازِلُونَ *nāzilūna*; رَجُلْ *raǧul* for رَجُلٌ *raǧulun*; مَرْحَبَا *marḥabā* for مَرْحَبًا *marḥaban*; فَاطِمَهْ *Fāṭimah* for فَاطِمَةُ.

11. *Numerical Signs and Abbreviations*. The usual Arabic cyphers are the following:

٠, ١, ٢, ٣, ٤, ٥, ٦, ٧, ٨, ٩
0, 1, 2, 3, 4, 5, 6, 7, 8, 9.

The tens, hundreds &c., are written to the left of the units &c. as ١٩ 19, ١٨٩٥ 1895.

The following are a few of the most frequently occurring abbreviations:

عم = عَلَيْهِ ٱلسَّلَامُ *'alaihi-ssalāmu* Peace be upon him! |

صلعم = صَلَّى ٱللّٰهُ عَلَيْهِ وَسَلَّمَ *salla-llāhu 'alaihi wasallama* God bless him and give him peace (said of Mohammed).

II. ETYMOLOGY (§§ 12—96).

Chapter I. The Pronoun. (§§ 12—15.)

The personal pronouns are either independent or suffixed. The *independent* or *separate* personal pronouns have the following forms: *a.*

	Sing.	Plur.	Dual
I. Pers.	أَنَا	نَحْنُ	
II. Pers. masc.	أَنْتَ	أَنْتُمْ (أَنْتُمُ)	أَنْتُمَا
II. Pers. fem.	أَنْتِ	أَنْتُنَّ	
III. Pers. masc.	هُوَ	هُمْ (هُمُ)	هُمَا
III. Pers. fem.	هِيَ	هُنَّ	

NOTE 1. The second syllable of the pronoun of the first pers. singular, although written with ا, is short. — The forms in parentheses (2nd and 3rd pers. plural) are used particularly before Waṣla (§ 6 *d*); these final vowels are originally long.

NOTE 2. When joined to وَ and فَ (see § 95) the pronouns of the 3rd pers. sing. may lose their first vowel e. g. فَهْىَ, وَهْوَ.

The *suffixed personal pronouns*, which joined to a   *b.* noun indicate the genitive, joined to a verb, the accusative, are the following:

12. THE PRONOUN.

		Sing.	Plur.	Dual
I. Pers.	with nouns	ـِي	ـنَا	
	with verbs	ـنِي		
II. Pers.	masc.	ـكَ	ـكُمْ	ـكُمَا
	fem.	ـكِ	ـكُنَّ	
III. Pers.	masc.	ـهُ	ـهُمْ	ـهُمَا
	fem.	ـهَا	ـهُنَّ	

c. Before a connective Alif (§ 6 d) the suffix pron. of the 1. pers. singular may receive as helping-vowel the *a* which belonged to it originally; thus we may write أَعْطَانِىَ ٱلْكِتَابَ or أَعْطَانِىَ ٱلْكِتَابَ. After *ā, ī* and *ai* the nominal suffix of the 1. pers. sing. has the form ىَ *ya*. Occasionally (in the Ḳur'ān particularly) the suffix of the 1. pers. sing. is indicated by a simple *i*, of which the sign is Kesr ـِ, as رَبِّ my lord! In the same way the corresponding verbal suffix may be only نِ *ni*.

d. After an immediately preceding *i* or *ai* the suffixes هُ, هُمَا, هُمْ, هُنَّ substitute the vowel *i* for *u*, thus assuming the forms هِ, هِنَّ, هِمْ, هِمَا; e. g. مَالِهِ instead of مَالِهُ. Before the connective Alif هِمْ generally becomes هِمِ. — The suffixes كُمْ and هُمْ resume their original forms كُمْ and هُمْ before a connective Alif.

For further information regarding the affixing of these pronominal forms see § 82 and the table of paradigms No. XXIII.

e. The reflexive pronoun, when carrying a certain amount of emphasis with it, is generally expressed by the word نَفْسٌ *nafsun* soul, to which the proper suffixes are appended. In many cases. however, the personal pronoun suffices to express the reflexive.

13. The *demonstrative pronouns* are the following (with their inflexion compare § 76 *a*).

a. The simple pronoun (rare)

	Masc.	Fem.
Sing.	ذَا	تَا ; تِهِ , تِي ; ذِهِ , ذِي
Dual Nom.	ذَانِ	تَانِ
Dual Gen. Acc.	ذَيْنِ	تَيْنِ
Plur.	أُولَى (*ŭlā*) or أُولَاءِ (*ŭlā'i*)	

b. This simple pronoun combines:

(1) with the demonstrative particle هَا, generally written defectively (هٰ or less correctly هٰ § 2*b*). The result is the usual demonstrative pronoun to indicate that which is near at hand (*this, these*):

13. THE DEMONSTRATIVE PRONOUNS.

	Masc.	Fem.
Sing.	هٰذَا	هٰذِهِ (هٰذِى)
Dual Nom.	هٰذَانِ	هٰتَانِ
Dual Gen. Acc.	هٰذَيْنِ	هٰتَيْنِ
Plur.	هٰؤُلَاءِ	

c. The simple demonstrative combines (2) with a suffix of the second person. Only in the older Arabic, particularly that of the Ḳur'ân, however, does the suffix vary according to the number of persons addressed (e. g. plur. ذٰلِكُمْ, dual ذٰلِكُمَا), elsewhere it appears uniformly as كَ. There is also a form with لِ before كَ. The result is two forms of the demonstrative pronoun to indicate that which is more remote (*that, those*):

	Masc.	Fem.
Sing.	تِلْكَ (تِيكَ), تَاكَ	ذَاكَ, ذٰلِكَ (ذَالِكَ, ذَلِكَ)
Dual Nom.	ذَانِكَ, ذَانِكَ	تَانِكَ, تَانِكَ
Dual Gen. Acc.	ذَيْنِكَ, ذَيْنِكَ	تَيْنِكَ, تَيْنِكَ
Plur.	أُولَائِكَ (أُولَاكَ), rarely أُولَالِكَ	

d. Among the demonstratives we must also place the article اَلْ (see § 5*b*). When the noun, in the circumstances detailed in § 6 ƒ 1, begins with a ل, this letter has a Tešdīd placed over it and the ل of the article is dropped. Thus we get لِلَّيْلَةِ for لِٱلَّيْلَةِ; so too لِلَّهِ for لِٱللَّهِ (§ 5 note).

The *relative pronouns* are the following: 14.

a. اَلَّذِى who, which, that,—originally a compound demonstrative with the article as one of its elements (hence the connective Alif)—declined as follows:

		Masc.	Fem.
Sing.		اَلَّذِى	اَلَّتِى
Dual	Nom.	اَللَّذَانِ	اَللَّتَانِ
	Gen. Acc.	اَللَّذَيْنِ	اَللَّتَيْنِ
Plur.		اَلَّذِينَ	اَللَّاتِى, اَللَّوَاتِى

b. مَنْ (indeclinable) one who, such (a one) as, he who, those who.

مَا (indeclinable) that which, something which.

c. Among the relative pronouns may also be included أَىٌّ, fem. أَيَّةٌ he who, she who. This word is declinable in the sing., but the masc. often takes the place of the fem. It also combines with the prons. in *b* above to form أَيُّمَنْ every one who, whosoever; and أَيُّمَا whatsoever.

15. The *interrogative pronouns* are:

مَنْ who?

مَا what? Frequently strengthened by the addition of the demonstrative ذَا: مَا ذَا what then?

أَيٌّ, fem. أَيَّةٌ what sort of? which?

NOTE. مَا after prepositions is shortened to مَ e. g. لِمَ why? With this interrogative مَا is also connected the interrogative particle كَمْ how much? ܟܡܐ ܣܓܝ ܐܢܘܢ

Chapter II. The Verb. (§§ 16—54.)

16.   The great majority of Arabic verbs have three radical letters; only a small minority have four radicals. The ground-form of verbs, according to which they are arranged in grammar and dictionary, is the third person singular of the perfect. The verb فَعَلَ (to do) is used as a model paradigm.

> NOTE. Since all Arabic dictionaries give the verbal and nominal derivatives under their respective root-forms, it is necessary, in order to find the three radicals with ease, to note carefully what consonants are employed in the formation of verbs and nouns as prefixes and affixes to, and as infixes in, the stem.

17.   From this ground-form or root, which is named by grammarians the first stem, other stems are derived by a series of uniform changes, represented by

modifications of the verb فَعَلَ, but usually referred to by their respective numbers in the series. Thus we speak of "the eighth stem", (indicated in the dictionary simply by VIII) not as in Hebrew and Syriac of the Piel, the Afel &c. The following stems, the order of which must be carefully noted, are those most frequently met with:

I فَعَلَ   IV أَفْعَلَ   VII اِنْفَعَلَ   X   اِسْتَفْعَلَ
II فَعَّلَ   V   تَفَعَّلَ   VIII اِفْتَعَلَ   XI اِفْعَالَّ
III فَاعَلَ   VI تَفَاعَلَ   IX   اِفْعَلَّ

NOTE a. Of these No. IX and especially No. XI are of less frequent occurrence; still more rare are XII اِفْعَوْعَلَ, XIII اِفْعَوَّلَ, XIV اِفْعَنْلَلَ, XV اِفْعَنْلَى. Which of these derived stems are formed from any given verb, and to what extent the meaning of the ground-form is modified by them, will be found in the dictionary under each verb.

NOTE b. In many cases the verb is used to express the idea that some one wishes to do something or has something done; thus قَتَلَهُ "he killed him" may also signify "he wished to kill him", and ضَرَبَ عُنُقَهُ "he cut off his head (prop. neck)" may mean "he had (*curavit*) his head cut off."

18. The ground-form I, in the majority of verbs, takes the form فَعَلَ, e. g. قَتَلَ to kill; there is also — mostly with intransitive verbs — a form فَعِلَ (cf. כָּבֵד), e. g. حَزِنَ to be sad, عَمِلَ to do (transitive), and also a

form فَعُلَ (cf. קָטֹן), confined to intransitive verbs, as حَسُنَ to be beautiful. Sometimes both the transitive and intransitive forms, فَعَلَ and فَعِلَ or فَعُلَ, are found side by side in the same verb. One and the same verb, again, may have both the forms فَعِلَ and فَعُلَ.

19. The II. stem فَعَّلَ (corresponding to the Hebrew Pi''ēl) usually denotes a greater intensity of the action expressed by the simple verb. This intensification may affect the subject, object or qualifying adjunct, as قَتَّلَ to kill many people, to massacre (intensification of the object). In the majority of verbs, however, the II. stem is causative as عَلِمَ to know, عَلَّمَ cause to know, to teach. It is also declarative—as in كَذَبَ to lie, كَذَّبَ to take one for, declare one to be, a liar—and denominative, as in جَيَّشَ to collect an army (جَيْش).

20. The III. stem فَاعَلَ expresses an attempt or effort to perform the action of the simple verb on some person, to influence some person or thing. Thus قَتَلَ to kill, but قَاتَلَ to try to kill, to fight with; كَتَبَ to write, كَاتَبَ to correspond with (with accusative of

the person corresponded with). This stem also means to exercise some abstract quality on a person or thing, e. g. لَانَ to be soft, gentle, لَايَنَ to exercise gentleness on some one, to treat one kindly.

21. The IV. stem أَفْعَلَ (the Hebrew Hiph'îl) has a causative signification, as صَلَحَ to be in good condition, أَصْلَحَ to bring into good condition. Very frequently we find, with this stem, denominative verbs which appear to us as intransitive, but to the Arab as possessing an implicit transitive force, and which express the idea of action in a certain definite direction, as أَحْسَنَ to do good. Frequently, too, verbs of this stem convey the idea of going to a place, of entering upon a certain period or condition; e. g. أَغْرَبَ to go towards the West, أَصْبَحَ to enter upon the period of the morning, to be something in the morning, أَشْرَفَ to reach the top, to be high; أَقَامَ (from قَامَ rise up, stand) to halt, to stay.

22. The V. stem تَفَعَّلَ (Hebrew Hithpa''ēl), a sort of middle voice is formed from the II. stem and has both a reflexive and a reciprocal meaning, e. g. تَكَبَّرَ to make one's self great, تَعَلَّمَ to let one's self be taught, to learn. Sometimes a verb in the V. stem conveys the

idea of giving one's self out as something, e. g. تَنَبَّأ
to give one's self out for, to conduct one's self as, a
prophet.

23. The VI. stem تَفَاعَلَ, derived from the III. stem, is
the reflexive form of the latter, and has a reflexive or
reciprocal signification, as تَجَاسَرَ to show one's self bold;
تَقَاتَلَ to fight one another (usually in the plural).
Another signification is seen, for example, in تَعَالَى, VI
form of عَلَا to be high, which means to exalt one's
self and then simply: to be exalted.

24. The VII. stem اِنْفَعَلَ (the Hebrew Niph'al with
the connective Alif acc. to § 6 a), derived in most cases
from the I. stem, is a middle or reflexive form of the
latter. Its signification may also be described as
quasi-passive, e. g. كَسَرَ to break اِنْكَسَرَ, to break or
be broken in pieces.

25. The VIII. stem اِفْتَعَلَ, (with connective Alif § 6 a)
is likewise a middle and reflexive form, for the most
part of the I. stem, as اِعْتَرَضَ, to oppose one's self,
object to; sometimes also with reciprocal signification
as اِخْتَصَمَ, to dispute, contend with each other.

NOTE. In the case of verbs whose first radical is ط, ض, ص
or ظ, the ت of the VIII. stem is changed to the emphatic ط, and

is even assimilated to the first radical, when that letter is a dental as اِصْطَبَغَ, instead of اِصْتَبَغَ from صَبَغَ; اِظَّلَمَ or اِظْطَلَمَ for اِظْتَلَمَ from ظَلَمَ; ت is sometimes assimilated also to a preceding ث, e. g. اِثَّبَتَ or اِثْتَبَتَ from ثَبَتَ properly اِثْتَبَتَ; after د, ذ and ز ت is changed into the soft د, e. g. اِدَّرَكَ for اِدْتَرَكَ; اِزْدَادَ from زَادَ for اِزْتَادَ.

The IX. stem اِفْعَلَّ (as also the XI. stem اِفْعَالَّ, 26. both with connective Alif) is <u>used of verbs</u> which denote the <u>possession of inherent qualities</u> such as colours or bodily defects, e. g. from the stem صفر: اِصْفَرَّ to <u>be or become yellow</u>; from the stem عور: اِعْوَرَّ to be one-eyed; from the stem حمر: اِحْمَارَّ to be red.

The X. stem اِسْتَفْعَلَ, (with connective Alif) is 27. primarily a reflexive of the IV. أَفْعَلَ (otherwise a reflexive, formed on the analogy of the VIII. stem, from a stem سَفْعَلَ with a prefixed s), as from the stem وحش IV. أَوْحَشَ to grieve: X. اِسْتَوْحَشَ to grieve (one's self). Very frequently the X. stem denotes also to wish or to <u>beg something for one's self</u>, e. g. from أَغْفَرَ to pardon, X.: اِسْتَغْفَرَ to ask for pardon; or to think that something is so, as وَجَبَ to be necessary, IV: أَوْجَبَ to make necessary, X: اِسْتَوْجَبَ to consider something as necessary for one's self.

28. The quadriliteral stems are denoted, for the verbal and nominal forms, by the paradigm فَعْلَلَ (that is by the addition of a fourth radical to فَعَلَ), and consist for the most part of two stems, of which the first may be said to correspond to the second stem of the triliteral verb (for فَعَّلَ is in reality فَعْعَلَ), and the second تَفَعْلَلَ to the fifth, e. g. كَبْكَبَ to overturn, cast down, تَكَبْكَبَ fall down.

NOTE. The stems III اِفْعَنْلَلَ and IV اِفْعَلَلَّ (the last corresponding to the IX. stem of the triliterals) are rare e. g. اِطْمَأَنَّ, to be quiet, from a stem طمأن.

29. In addition to the *active*, the Arabic verb has a *passive* voice. This passive is formed in the perfect in such a way that in place of the *a*-vowels of the active we have the order *u-i-a* (*i* with the second, *a* with the third radical); thus the act. of stem I. is فَعَلَ, the pass: فُعِلَ. The additional formative syllables of the derived stems also receive the vowel *u*, e. g. pass. V. تُفُعِّلَ, VIII أُفْتُعِلَ (with connective Alif).

30. The Arabic verb has two principal tenses, a *perfect*
a. which, generally speaking, denotes a completed action, and an *imperfect* which in general denotes an uncompleted action.

b. The imperfect is formed by adding the prefix يَ *ya* for the active of the I., V., VI., VII., VIII., IX. and X. stems, and the prefix يُ *yu* for the active of the II., III. und IV. stems, and for the passive of all the stems without exception.

c. In the case of verbs of which فَعَلَ is the type, the second radical, in the impf. act. of stem I., may receive one or other of the vowels *u, i, a*. Which of the three must be used for a particular verb will be found indicated in the dictionary under that verb (e. g. قتل impf. *u*) and should be taken careful note of. Those verbs, on the other hand, of which فَعِلَ (with *i*-vowel) is the type, together with all passives point their second radical with *a* only, thus impf. act. I. يَفْعَلَ; pass. يُفْعَلَ. Those verbs, finally, of which فَعُلَ (with *u*-vowel) is the type, take *u* with the second radical for the imperfect. As regards the active imperfect of the derived stems, the second radical takes *i* throughout, with the exception of stems V. and VI. where it takes *a*; thus impf. II. يُفَعِّلَ but V. يَتَفَعَّلَ.

31. In the imperfect various *Moods* are distinguished, namely the ordinary mood which we call the *indicative*, the dependent mood or *subjunctive*, and a *modus apocopatus* (sometimes called the jussive). These are distinguished as follows: in the indicative the last radical, when final, always takes *u*, as impf. I يَفْعَلُ III. يُفَاعِلُ; in the subjunctive always *a*, as يَفْعَلَ while in the *apocopatus* the third radical is vowelless. In addition to the above there is a double modus energicus, which is formed by appending the syllables *anna* or *an* (in some forms only *n*) to the impf. as يَفْعَلَنْ or يَفْعَلَنَّ.

NOTE. As the modus energicus is of comparatively rare occurrence, it is given in the tables of paradigms only in the case of the ordinary strong verb. From the examples there given it may easily be formed for the other verbs.

32. The *imperative* agrees with the apocopated imperfect as regards vocalisation and termination, except that the prefixes *ya* or *yu* are wanting. In the imper. of the I. stem a helping vowel (therefore with connective Alif § 6 *a*) is prefixed in all cases where the first consonant is without a vowel of its own. This vowel disappears, however, in pronunciation as soon as the word ceases to stand alone, e. g. اِفْعَلْ but قَالَ آخْرُجْ. The same applies to stems VII.—X. The imperative has the same energetic bye-forms as the imperfect.

NOTE. In the imper. of stem I the prosthetic vowel is *u* when the second radical has *u*, as اُقْتُلْ, but *i* when it is pointed with *a* or *i*, as اِزْبِنْ, اِفْعَلْ.

NOTE b. In the imper. of the IV. stem the prosthetic ﺍ, which is characteristic of the stem, is retained, although it disappears after the prefixed ﻳ of the impf. Hence impf. يُفْعَلُ (for يُأَفْعَلُ), but imper. أَفْعِلْ.

In the perfect, imperfect and imperative, there **33.** are, in addition to the singular and plural, dual forms for the second and third persons. Verbs are inflected by the addition of modified and abbreviated forms of the personal pronouns, and of the dual and plural terminations of nouns, to the ground-forms فَعَلَ and يَفْعَلُ (for the terminations *āni* and *ūna* of the impf. indic. vid. § 76 a). The terminations just named, along with the ending *īna* of the 2. pers. fem. sing., drop the syllables *ni* and *na* in the subjunctive, the apocopated imperfect and the imperative. The ﺍ, which appears in the paradigm after the final ‍ُو in the perf. and in these shortened forms of the impf. and imper., has no phonetic value (cf. § 2 e).

As to the *prefixes* of the impf., it is to be noted that in place of the prefix ﻳ of the 3. pers. masc., we have ﺗ as the prefix of the 2. pers. sing. and plur., and of the 3 pers. fem. of the sing., ﺍ to indicate the 1. pers. sing., and ﻧ the 1. pers. plur.

The *affixes* employed in the inflexion of the verb are given in paradigm I.

Note a. In the V. and VI. forms of verbs whose first letter is a dental or a sibilant, the formative prefix occasionally drops its vowel and is assimilated to the first radical of the verb, in which case the perf. and imper. have a helping vowel (§ 6) prefixed e. g. اِدَّثَّرَ wrap one's self up, impf. يَدَّثَّرُ.

Note b. In the impf. of these two stems, the prefix ت may be treated in such a way that instead of the two syllables تَتَ only ت remains, e. g. from قتل 2. pers. msc. impf. V. تَتَقَتَّلُ for تَتَتَقَتَّلُ.

Note c. In the impf. VII. and VIII. stems the tone remains on the same syllable on which it falls in the perf., contrary to the rule laid down in § 9; thus يَنْقَتِلُ يَقْتَتِلُ *yanḳátilu*, *yaḳtátilu*.

For the conjugation of the strong verb with three radicals see paradigms II, III and V, for that of the quadriliteral verbs see paradigm IV. In the paradigms the participles and infinitives are also given, although the discussion of these forms has been deferred to §§ 60 and 61.

34. Among the ordinary strong verbs must also be reckoned the so-called verbs *mediae geminatae*, i. e. verbs whose second and third radicals are identical.

A contraction of these last two radicals takes place in all those cases in which

a. 1) the first, second and third radicals have each a short vowel; in this case the vowel of the second radical is always dropped, e. g. فَرَّ (to flee) contracted from فَرَرَ (which statement is not to be understood as implying that a form فَرَرَ once really existed in Arabic) 3. p. perf. pass. I. فُرَّ from فُرِرَ; 3. p. impf. VII. يَنْفَرُّ from يَنْفَرِرُ;

2) When the first two radicals have each a short, *b.*
and the third a long, vowel, e. g. 3. p. dual masc.
perf. فَرَّا from فَرَرَا;

3) Generally also when the first radical has a long *c.*
ā, e. g. 3. s. m. perf. of the III. stem فَارَّ contracted from
فَارَرَ (which is also found), passive فُورِرَ.

When the first radical is vowelless and the second **35.**
has a short vowel, then contraction takes place and
the vowel of the second radical passes over to the
first. Thus 3. pers. impf. act. يَفِرّ for يَفْرِرُ; pass. يُفَرّ
from يُفْرَرُ.

When the third radical is vowelless, there is no **36.**
contraction in the body of the word: e. g. 2. pers. sing.
masc. perf. act. فَرَرْتَ; 3. pers. plur. fem. impf. act.
يَفْرُرْنَ. But when the third radical stands at the end
of a verbal form with no vowel following, as in various
forms of the apocopated impf. and the 2. pers. sing.
masc. of the imper., we find the full forms اُفْرُرْ, يَفْرُرْ
only in the dialects. As a rule contraction takes place
and an additional vowel is assumed at the end in
order to preserve the doubling of the radical; thus
we have فِرَّ, يَفِرَّ, from فَرَّ imper. فُرَّ.

3*

NOTE. In the case of verbs of the forms فَعِلَ and فَعُلَ the vowel of the second radical appears only in the uncontracted form e. g. مَلَّ to loathe, 1. pers. perf. مَلِلْتُ; hence the vowel *a* of the impf. يَمَلُّ.

For the conjugation of verbs *mediae geminatae* see paradigms Nos. VI—VIII; model verb فَرَّ to flee.

37.   Those verbs that have a Hamza ء as first, second or third radical are for the most part regular, as أَثَرَ to make an impression, impf. يَأْثِرُ; قَرَأَ to read, impf. يَقْرَأُ. In certain cases we find, according to § 4 *b*, و or ى (without points) as bearers of the Hamza, or ء may stand without a bearer, thus 3. s. m. perf. act. كَئِبَ to be sad, بَوُسَ to be brave; 3. s. m. impf. passive of أَثَرَ: يُوثَرُ; 3. sing. masc. perf. act. خَطِئَ to err, fem. خَطِئَتْ; 3. s. m. impf. act. of سَأَلَ to ask: يَسْأَلُ. Occasionally an ا takes the place of two Alifs, according to § 7; e. g. 3. s. m. perf. III. of أَثَرَ: آثَرَ for أَأْثَرَ; VI. of لَأَمَ (bind up a wound &c.) تَلَاءَمَ.

38.   While in all these cases the ء may easily be distinguished as the third radical of the verb, there are a few forms in which the verba hamzata are more difficult to distinguish, inasmuch as the ء sometimes entirely disappears; from this point of view these verbs ought rather to be reckoned among the weak verbs (§ 39 ff.). The most important of such cases are the following:

1) After اَ, اُ, اِ (also after a connective Alif اَ, اُ, اِ) ء gives up its power as a consonant (cf. § 7); hence, in place of 'a', 'u', 'i' simply 'ā, 'ū, 'ī, e. g. 3. s. m. perf. IV. of اَأْثَرَ: آثَرَ for اُأْثِرَ; 3. s. m. perf. pass. IV. of اَأْثَرَ is اُوثِرَ in place of اُأْثِرَ. So also imper. I. اِيثِرْ for اِأْثِرْ. *a.*

2) In the imper. of the I. form the verbs أَخَذَ take, أَكَلَ eat, أَمَرَ order, drop the ء altogether: خُذْ, كُلْ, مُرْ; in the same way, from سَأَلَ to ask, the imperative is either اِسْأَلْ or سَلْ &c. *b.*

NOTE. Should وَ or فَ come to stand as inseparable particles (§ 87) before one of the imperatives under *a*, the prosthetic Alif is dropped and the radical Hamza reappears, receiving, as its bearer, an Alif on account of the preceding Fath, as in فَأْتِ. The same holds good in the case of two separate words: thus 3. s. m. perf. pass. VIII of أَمِنَ connected with a preceding word becomes اَلَّذِى أُؤْتُمِنَ elladi-'tumina.

3) In the VI. form the ء of verbs primae ء is sometimes changed to و, as تَوَامَرَ in place of تَآمَرَ (for تَأَامَرَ). *c.*

4) In the VIII. form the ء of the verb أَخَذَ is *d*

assimilated to the following ت, the result being ت, as اِتَّخَذَ instead of an original اِئْتَخَذَ, impf. يَتَّخِذُ, but from أَمَرَ, to order, اِيتَمَرَ.

For the conjugation of the verba hamzata see paradigm IX.

The Weak Verbs.

39. The weak verbal stems are those having a و, or a ى as first, second or third radical; under inflection these semivowels in some cases resolve themselves into full vowels, in others they are treated as consonants.

40. The *Verbs primae* و *and* ى differ from the strong verbs in the following points:

a. 1) In the impf. and imper. of the I stem a number of verbs primae و surrender their first radical and take the vowel *i* with their second (cf. ילד), as وَلَدَ to bring forth, impf. يَلِدُ, imper. لِدْ.

b. 2) Under the influence of a guttural a few verbs take *a* in place of *i* with their second radical, dropping the و, however, like the others, as وَضَعَ to lay, impf. يَضَعُ; so وَقَعَ to fall, وَهَبَ to give and others (see the dictionaries).

c. 3) In verbs primae ى, يِــْـ is changed to *ū*, e. g. the impf. IV of يَقِظَ to be awake, properly يُيْقِظْ, becomes يُوقِظْ.

4) In the VIII. stem the first radical of verbs primae و and ى is assimilated to the following ت, e. g. from وَعَدَ to promise, اِتَّعَدَ for اِوْتَعَدَ (cf. § 38 d).

NOTE. A few verbs of the form فَعَلَ also give up the first radical in the imperf. as وَرِثَ to inherit, impf. يَرِثُ (cf. § 18).

For the conjugation of the verbs primae و and ى see paradigm X where will be found the principal forms of the verbs وَصَلَ to arrive, وَدَعَ to leave, وَسِخَ to be dirty, وَجِلَ to be anxious, وَسِنَ to be sleepy, يَسُرَ to be easy.

41. *Verbs mediae* و *and* ى. In the II., III., V., VI. and IX. stems, و and ى are treated as consonants, and the inflexion is the same as that of the strong verb; thus 3. s. m. perf. II of قَالَ (to say) med. و: قَوَّلَ, 3. s. m. perf. III of سَارَ (to travel) med. ى: سَايَرَ. In the other stems these verbs are inflected according to the following rules:

42. Long *ā* takes the place of the middle radical:

a. in the perf. active of the I., IV., VII., VIII. and X. stems, as قَالَ, أَقَالَ, اِنْقَالَ, اِقْتَالَ, اِسْتَقَالَ;

b. in the impf. passive of the same stems, as يُقَالُ, يُنْقَالُ, يُقْتَالُ, يُسْتَقَالُ;

c. in the impf. active of VII. and VIII., as يَنْقَالُ, يَقْتَالُ;

d. in the impf. active of the I. stem of verbs of the form فَعِلَ e. g. خَافَ to fear, impf. يَخَافُ.

43. Long *ī* takes the place of the middle radical:

a. in the perf. passive of the I., IV., VII., VIII. and X. stems as قِيلَ, أُقِيلَ, اُنْقِيلَ, اُقْتِيلَ, اُسْتُقِيلَ;

b. in the impf. active of IV. und X., as يُقِيلُ, يَسْتَقِيلُ;

c. in the impf. active of verbs med. ى, as يَسِيرُ. The corresponding form of verbs med. و, on the other hand, takes long *ū*, as يَقُولُ.

NOTE. The nature of the phonetic changes just detailed will be more readily understood from the standpoint of the strong verb if it be noted that َـَـْ, َـِـْ, َـُـْ, ـِـَـْ; ـَـُـْ, ـِـُـْ pass into *ā*; ـَـِـْ, ـُـِـْ, ـِـَـْ, ـِـُـْ into *ī*; ـَـُـْ into *ū*. It is not meant by this that the corresponding strong forms were ever really found, in these verbs, at any period of the language.

44. The whole of the long vowels mentioned in §§ 42—43 are shortened (§ 8) in a shut syllable, e. g.:

2. s. m. perf. act. IV. of قَالَ and سَارَ: أَقَلْتُ and أَسَرْتُ;

3. sing. masc. apoc. impf. pass. I يُقَلْ, يُسَرْ (with the tone on the last syllable as if contravening § 9).

2. pers. masc. sing. imper. I. of خَافَ (§ 42 *d*): خَفْ (but plur. خَافُوا);

2. pers. masc. sing. perf. pass. قِلْتَ;

3. pers. sing. masc. apoc. impf. act. IV. يُقِلْ;

2. pers. sing. masc. imper. I: قُلْ, سِرْ.

In the perf. active of I, verbs med. و take ŭ where we should expect ă, (cf. קָמְתָּ) as قُلْتَ, while verbs med. ي take ĭ, as سِرْتَ; ĭ is also found in verbs of the form فَعِلَ, as خِفْتَ from خَافَ (for a theoretical خَوِفَ).

NOTE a. Instead of the apocop. impf. يَكُنْ &c. from كَانَ, to be, we sometimes find the still shorter form يَكُ.

NOTE b. From a few verbs med. و and ي strong forms are found in stems I., IV., VIII., X.; e. g. IV. أَحْوَجَ compel; X. اِسْتَصْوَبَ to find correct, a denominative form from صَوَابٌ correct.

For the conjugation of these verbs see paradigms XI—XIV.

Verbs ultimae و *and* ي. Verbs ultimae و pass into 45. ultimae ي in all the derived stems, and in the perf. and impf. passive of the I stem; thus from غزو we have 3 s. m. perf. II غَزَّى. The same applies to the active of stem I of the form فَعِلَ; thus رَضِوَ becomes رَضِىَ (to have pleasure in).

If the second radical has ă, this vowel is changed 46. in every case into a long final ā. In order to distinguish a. the stems ult. ي from those ult. و, this final ā is in the former case indicated by ي, in the latter by ا (this applies only to the 3. s. m. perf. act. I). Thus رَمَى to throw, غَزَا carry on a war; but II. رَمَّى, غَزَّى &c. Similarly in the imperfects (cf. § 45), e. g. indic. and subj. pass. II يُرَمَّى (in place of a theoretical يُرْمَى and يُرْمَىَ); impf. act. I of رَضِىَ, يَرْضَى; impf. act. V. يَتَرَمَّى.

NOTE. With the same reservation as under § 43c note, we would call attention to the fact that the combinations ـَـَى, ـَـَوْ, ـُـَوْ, ـِـَى all pass into long *ā*.

b. In all the cases mentioned in the preceding subsection, a diphthong (§ 2 *a*) appears before the inflectional additions that begin with a consonant. Thus: 2. sing. masc. perf. act. I رَمَيْتَ; from غَزَا: غَزَوْتَ; II غَزَّيْتَ, رَمَّيْتَ &c.

c. In the case also of the inflectional additions *ū*, *ūna*, *īna* (and its shortened form *ī*), the *a* of the second radical, (after the elision of the third radical) unites with their initial vowel to form a diphthong. Thus: 3. pers. masc. plur. perf. act. I. غَزَوْا, رَمَوْا, do. impf. pass. II. يُرَمَّوْا, subj. يُرَمَّوْا; do. act. I. يَرْضَوْنَ, V. يَتَرَمَّوْنَ; 2. pers. fem. sing. of the last يَتَرَمَّيْنَ, subj. يَتَرَمَّى.

d. Before the dual terminations *ā* and *āni* the last radical of this class of verbs is treated as a strong letter, e. g. 3. pers. perf. act. I. غَزَوَا, رَمَيَا; impf. pass. II. يُرَمَّيَانِ &c. By the addition of the termination *at*,

47. VERBS ULTIMAE و AND ي.

the 3. pers. fem. sing. of the perfect must originally have ended in *āt*; this ending, however, has now become *at* in accordance with § 8, as غَزَتْ; رَمَتْ. According to the analogy of the above is also formed the 3. pers. fem. of the dual; thus we find رَمَتَا, غَزَتَا (where we should expect غَزَاتَا, رَمَاتَا).

In the impf. active of stem I, verbs ult. و of the form فَعَلَ take an *u*, those ult. ي an *i*, the third radical quiescing in these vowels. The ending *u* of the imperf. is lost, e. g. يَرْمِي, يَغْزُو. The imperfects active of the derived forms (with the exception of V and VI) are formed on the model of the last mentioned forms, as II يَغَزِّي, يَرَمِّي and so on. *a.*

NOTE. With the same reservation as under § 43c note, it may be pointed out that ـُو passes into *ū*, ـِي into *ī*.

Affixes beginning with a consonant are appended *b.* in every case to the *ī* or the *ū* just mentioned, as 3. pers. fem. plur. impf. I. يَغَزُونَ, يَرْمِينَ; similarly in the perf., e. g. 2. sing. masc. perf. pass. رُمِيتَ; do. from فَعِلَ I. رَضِيتَ; from فَعُلَ I. سَرُوتَ.

If the second radical has *i* or *u*, the third radical *c.* is dropped and the terminations *ū*, *ūna*, *īna* added to the second, e. g. 3. plur. masc. perf. pass. رُمُوا (not

44   48. VERBS ULTIMAE و AND ى.   49. DOUBLY WEAK VERBS.

يَغْزُونَ، يَرْمُونَ 3. plur. masc. impf. act. ;غَزَوْا، (رَمِيُوا
(not يَغْزُوُونَ، يَرْمِيُونَ); 2. pers. fem. sing. impf.
تَغْزِينَ، تَرْمِينَ.

d. Before the dual endings *ā* and *āni*, as also before the terminations *a* of the 3. sing. masc. perf., *at* of the 3. sing. fem. perf., *atā* of the 3. fem. dual perf., and *a* of the subjunctive, the third radical is treated as a strong letter, if the second has *i* or *u*. Exx: 3. pers. masc. perf. act. سَرُوَ، رَضِىَ; do. pass. رُمِىَ ;غُزِىَ 3. pers. fem. perf. رَضِيَتْ، سَرُوَتْ; 3. pers. masc. dual رَضِيَا; fem. رَضِيَتَا; 3. pers. subj. act. I يَغْزُوَ، يَرْمِىَ; 3. pers. dual impf. يَغْزُوَانِ، يَرْمِيَانِ.

48.   In the apocopated impf. and in the imper. every final *ā*, *ī* and *ū* is shortened, as 3. pers. sing. masc. apoc. impf. يَغْزُ، يَرْمِ، يَرْضَ; 2. imper. اُغْزُ، اِرْمِ، اِرْضَ.

For the conjugation of these verbs see paradigms XV—XIX where various forms are given of the verbs غَزَا to carry on war, رَمَى to throw, رَضِىَ to be content, قَضَى to carry out, accomplish.

49.   Of verbs doubly weak the following are the principal varieties:

a.   Verbs primae و and ultimae ى, as وَقَى to take care of; impf. according to §§ 40 and 47 يَقِى, apoc. يَقِ.

50. THE VERB لَيْسَ.

The imper. is properly رِ, for which, however, when the word stands alone, i. e. in pause, we write رِهْ.

The verb رَأَى to see, which in the impf. elides *b.* the Hamza, throwing back its vowel *a* to the first radical. Thus يَرَى *yarā* for يَرْأَى *yar'ā*; 3. pers. pl. يَرَوْنَ; imper. رَ (acc. to *a* رَهْ), fem. رَىْ. The IV. form in the sense of 'to show' is similarly inflected: أَرَى for أَرْأَى, impf. يُرِى for يُرْئِى; perf. pass. أُرِىَ for أُرْئِىَ and so on.

The verb حَىَّ to live, properly حَيِىَ; impf. يَحْيَا (cf. *c.* § 2 *d* note) like a verb ult. ى or يَحَىُّ like a verb mediae geminatae; perf. IV أَحْيَا, perf. X اِسْتَحْيَى or اِسْتَحْيَا also contracted اِسْتَحَى (be ashamed).

The verb لَيْسَ 'there is not' (compounded of the 50. negative لَا and an obsolete Arabic noun corresponding to the Hebrew יֵשׁ) is inflected as follows:

	Sing.	Dual	Plural
3. masc.	لَيْسَ	لَيْسَا	لَيْسُوا
3. fem.	لَيْسَتْ	لَيْسَتَا	لَسْنَ
2. masc.	لَسْتَ	لَسْتُمَا	لَسْتُمْ
2. fem.	لَسْتِ		لَسْتُنَّ
1. com.	لَسْتُ		لَسْنَا

51. The verbs of praise and blame, نِعْمَ to be good and بِئْسَ to be bad, which are rarely conjugated, are written as above.

52. The Arab grammarians adduce as special forms the so-called *admirative* forms, that is, forms expressive of admiration. These are strictly the 3. s. m. perf. and 2. pers. imper. of the IV. stem, but have assumed a special signification; so مَا أَفْضَلَ زَيْدًا properly 'what has made Zaid excellent', and أَفْضِلْ بِزَيْدٍ prop. 'make Zaid excellent' both mean: how excellent is Zaid! — The verbs mediæ و and ى may in these forms take the inflection of the strong stems (§ 44 note *b*) as مَا أَهْوَنَ هٰذَا how easy this is!

53. The addition of the *pronominal suffixes* (§ 11 *b*) alters the form of the verb only to a slight extent.

a. The 2. pers. fem. sing. perf. with a suffix receives a long final vowel as ضَرَبْتِينِي.

b. The ا, standing after وُ‎ *ū* (§ 2 *e*), is dropped as قَتَلُوهُ from قَتَلُوا with the suff. of the 3. pers. sing. masc.

c. The ending تُمْ of the 2. pers. pl. perf. becomes تُمُو (cf. § 12 *a*, note 1), as قَتَلْتُمُونِي from قَتَلْتُمْ with the suff. of the 1. pers. sing.

d. Before the suffixes to the 1. pers. sing. and plur.,

ىِ and اَن, the final *na* of the 2. fem. sing. and 3. and 2. masc. plur. impf. is sometimes dropped (so that these forms become identical with those of the subjunctive and apocopated moods). Ex.: تَضْرِبِينِى alongside of the more common تَضْرِبِينَنِى thou (fem.) strikest me; يَضْرِبُونَا alongside of the more common يَضْرِبُونَنَا they strike us.

54. When the object of an active verb consists of a personal pronoun, and this object is, for the sake of emphasis, made to precede the verb, then instead of the ordinary suffixes appended to the verb the sign of the accusative إِيَّا (אֶת, אֵת) is employed with the suffixes of the noun (with the suff. of 1. pers. sing. إِيَّاىَ); e. g. إِيَّاكَ نَعْبُدُ to *thee* we pray.

b. The Arabic verb may have two suffixes appended at the same time, in which case the pronoun of the 1. person precedes those of the 2. and 3. persons, and the pronoun of the 2. person that of the third, as أَعْطَانِيهِ he gave it me; frequently, however, in place of the second suffix—more particularly when both pronouns are of the third person — we find the above mentioned periphrasis with إِيَّا as زَوَّجَهُ إِيَّاهَا he married him to her.

Chapter III. The Noun. (§§ 55—90).

a. *The Formation of Nouns.*

55. Nouns in the wider sense comprise 1) substantives, 2) adjectives, 3) numerals (§§ 91—93), and 4) pronouns (§§ 12—14). The noun, in the narrower sense, is limited to substantives and adjectives.

Primitive substantives is the name given to such substantives as cannot be derived from a verb. According to the usual arrangement of Arabic dictionaries, it is true, the primitive noun رَأْس, head (*un* affix) for example, is found under the verb رَأَس, but this verb is in all its significations denominative. On the other hand, it may fairly be maintained that a noun like رَأْس goes back to a hypothetical triliteral root ر + أ + س. — In contrast to these primitive nouns, we find a large number of nouns which are derived either from verbs or from other nouns, that is, which are either *deverbals* or *denominatives*. All the forms of the noun are indicated by paradigms from the root فعل (cf. § 15 ff.); thus we say of رَأْس as 'of the deverbal infinitive قَتْل killing, that it has the form فَعْل.

NOTE. The numerous foreign words which have found their way into Arabic, adapted from Persian and Aramaic, and indirectly from Greek and Latin, have also, to some extent, been reduced to Arabic nominal forms.

56, 57. THE FORMATION OF THE NOUN.

56. A number of nouns do not show the full complement
of (three) consonants (see §§ 16 and 90), as دَمٌ blood;
with the feminine termination (§ 73): أَمَةٌ a slave-girl;
to this group belong also nouns with a prefixed vowel
(connective Alif) as اِسْمٌ name, which accordingly must
be sought for in the dictionary under س.

b. Extremely common are the nominal forms with
one short vowel, like فَعْلٌ, فِعْلٌ, فُعْلٌ, e. g. رِجْلٌ foot,
according to the form فِعْلٌ. There are also nominal
forms with *two short* vowels: فَعَلٌ, فَعِلٌ, فُعُلٌ, فِعَلٌ,
فُعَلٌ, فُعُلٌ, e. g. رَجُلٌ a man, NF. فُعَلٌ; كِبَرٌ old age
NF. فِعَلٌ.

c. Next in order we may put nominal forms with a
long vowel either with the first radical فَاعِلٌ or with
the second فَعَالٌ, فِعَالٌ, فُعَالٌ, فَعُولٌ, فِعِيلٌ, or
with both فَاعُولٌ.

d. Nominal forms with doubling of the second radical
are such as حِمِّصٌ chick-pea NF. فِعِّلٌ; فَعَّالٌ (§ 63 a);
فِعِّيلٌ.

NOTE. By their mode of formation these nouns have been
raised to the rank of quadriliterals like those in §§ 57—58.

57. The *preformatives* employed in the formation of

nouns are the following (whose vowels vary according to circumstances): *a)* مُ cf. §§ 60 and 64. *b)* ة cf. § 61. *c)* ي as يَنْفُورٌ fugitive NF. يَفْعُولٌ from نَفَرَ to flee. *d)* ا (cf. §§ 62 *c*; 63 *b*), e. g. أُحْدُوثَةٌ story NF. أُفْعُولَةٌ from the stem حدث.

58. The *afformatives* or formative additions used in the formation of nouns are: *a)* ى— and آءُ— (see § 74). *b)* انٌ— (for substantives) or انٌ— (often to form adjectives) e. g. خَفَقَانٌ palpitation of the heart NF. فَعَلَانٌ from خَفَقَ; سَكْرَانُ drunk NF. فَعْلَانُ from سَكِرَ. *c)* —وتٌ (not originally Arabic) as مَلَكُوتٌ kingdom NF. فَعَلُوتٌ, which takes the masc. gend. in Arabic.

59. The *quadriliteral* nouns are denoted by the paradigm فعلل (§ 28) as عَقْرَبٌ scorpion NF. فَعْلَلٌ; صُنْدُوقٌ box NF. فُعْلُولٌ; مُعَسْكَرٌ military camp NF. مُفَعْلَلٌ; خُنْفَسَآءُ a species of beetle NF. فُعْلَلَآءُ.

60. From among the rich growth of nominal forms in Arabic a few deverbals and denominatives may be singled out for special attention. Such, of the former class, are the participles and infinitives, whose forms will be found among the paradigms of the verb.

The *participles* — the active is generally named a. nomen agentis, the passive nomen patientis — take the form فَاعِل for the active of the I stem, and for the passive the form مَفْعُول. In all the derived stems the participle is formed by prefixing the syllable مُ; in the active the second radical takes *i*, in the passive *a* (see below). As a rule, however, the active and passive participles of the derived stems take the vowels of the active and passive imperfs. with the exception of stems V and VI.

In addition to the participles there is a class of b. so-called *verbal adjectives,* which are in part treated as participles; they might be called quasi-participles, as حَسَن beautiful, from حَسُنَ.

The Arabic participles do not in themselves convey c. any suggestion of time; hence قَاتِل, for example, may mean 'one who has killed' as well as 'one who is killing', مَقْتُول 'one who ought to be killed' i. e. interficiendus as well as interfectus.

The Infinitive (nomen verbi) assumes various forms 61. in the I stem, and is therefore specially noted in the a. dictionaries under each verb. One of the most common forms is فَعْل, as قَتْل killing. The infinitives of فَعِلَ

4*

verbs (§ 28), as a rule, take the form فَعَل, e. g. from غَضِبَ, غَضَب the being angry. فُعُول and فَعَال are also common forms from intransitive verbs, as جُلُوس a sitting, from جَلَسَ; سَلَام health, from سَلِمَ. Infinitives are also found with the prefix *ma*, as دُخُول or مَدْخَل (for the same verb has frequently more than one form of the infinitive, sometimes with different meanings) from دَخَلَ to enter.

b. The infinitive of the II. stem has the form تَفْعِيل or تَفْعِلَة (cf. § 57*b*); the inf. of the III. stem the form فِعَال or مُفَاعَلَة (which last is identical with the fem. of the passive participle). The infinitives of IV., VII., VIII., IX. and X. are formed by the insertion of a long *ā* before the last radical; before this *ā* every short *ă* of the perf. becomes *ĭ*, as in the IV. stem إِنْعَال. The infinitives of V. and VI. take *u* after the second radical, as V. تَفَعُّل.

c. The Arabic infinitives do not contain the idea of time and may be used both in an active and in a passive sense. Thus قَتْل denotes the circumstance that some one has killed or has been killed, the idea of killing or of being killed.

62. VERBAL ADJECTIVES.

Synopsis of participles and infinitives

	Partcp. Act.	Partcp. Pass.	Infin.
I.	فَاعِلٌ	مَفْعُولٌ	cf. § 61 a
II.	مُفَعِّلٌ	مُفَعَّلٌ	تَفْعِيلٌ تَفْعِلَةٌ
III.	مُفَاعِلٌ	مُفَاعَلٌ	فِعَالٌ مُفَاعَلَةٌ
IV.	مُفْعِلٌ	مُفْعَلٌ	إِفْعَالٌ
V.	مُتَفَعِّلٌ	مُتَفَعَّلٌ	تَفَعُّلٌ
VI.	مُتَفَاعِلٌ	مُتَفَاعَلٌ	تَفَاعُلٌ
VII.	مُنْفَعِلٌ	مُنْفَعَلٌ	اِنْفِعَالٌ
VIII.	مُفْتَعِلٌ	مُفْتَعَلٌ	اِفْتِعَالٌ
IX.	مُفْعَلٌّ	— —	اِفْعِلَالٌ
X.	مُسْتَفْعِلٌ	مُسْتَفْعَلٌ	اِسْتِفْعَالٌ
Quadr. I.	مُفَعْلِلٌ	مُفَعْلَلٌ	فَعْلَلَةٌ فِعْلَالٌ
II.	مُتَفَعْلِلٌ	مُتَفَعْلَلٌ	تَفَعْلُلٌ

62. As regards *Verbal Adjectives* (cf. § 60 c), the following forms may be specially noted:

a. The form فَعِيلٌ, which occurs in both an active and a passive sense; as قَتِيلٌ killed, شَهِيدٌ a witness,

خَصِيم one who disputes with another (in the sense of مُخَاصِم part. act. of III).

b. فَعُول, e. g. كَذُوب (often an intensive form) given to lying.

c. أَفْعَل, a form denoting colours and physical defects, as أَصْفَر yellow; أَعْرَج lame; أَعْوَر (with و as a strong letter) one-eyed. For the formation of the feminine, see § 74 b.

63. Arabic has the means of expressing a heightened or intensive form of the root idea. Of such intensive forms the following are examples:

a. فَعَّال intensive form of فَاعِل and other verbal adjectives, as كَذَّاب (habitually) given to lying. As a denominative this form is in frequent use to denote trades or professions (nomina opificum) as خَبَّاز baker from خُبْز bread.

b. Very frequently there is derived from adjectives the form أَفْعَل in the sense of an elative (generally so named because including both comparative and superlative), as حَسَن beautiful, elative: أَحْسَن more b., most b.; صَغِير small, young, elative: أَصْغَر smaller, younger; smallest, youngest. The elatives, when standing in the predicate, do not admit of inflection for gender and number, as هُمْ أَفْضَلُ ٱلنَّاسِ they are the most excellent of men. When used in a comparative sense, they are mostly undetermined (§ 76 bc), and are followed by the preposition مِنْ in the sense of our "than" (properly 'at a distance from', 'measured from'). Used as superlatives, on the other hand, they are generally determined. For the feminine formation see § 74 b.

NOTE. No special elative is formed from the words خَيْر good and شَرّ bad, which are used as elatives in the form just given. As a matter of fact, the positive of other adjectives as well must sometimes be rendered by our superlative; thus كَبِيرُ ٱلنَّاسِ signifies the (absolutely) greatest of men.

To the class of deverbal nouns belong further: **64.**
Nouns of place and time formed with the prefix *a.*
مَ *ma,* as مَكْتَب the place where one writes, the school; also with the fem. termination as مَقْبَرَة a buryingplace.

NOTE. Nouns of place and time from the derived stems take the form of the pass. participle, as مُخْرَج (from the IV. stem of خَرَج to go out, of which IV. أَخْرَج caus.) the place to which or the time at which something is brought out; مُتَوَضَّأ (from V. stem) the place where the ritual washing is performed.

Nomina instrumenti, formed with the prefix مِ *mi, b.* as مِحْلَب milk-pail, from حَلَب to milk; مِفْتَاح key, from فَتَح to open.

c. Nomina speciei of the form فِعْلَة, as كِتْبَة the manner of writing, one's "calligraphy".

65. To the class of denominatives belong especially the nouns of relation and the diminutives.

a. By means of the termination ـِيّ (corresponding to the Hebrew ‿ִי, fem. ‿ִיָּה and ‿ִית) there is derived from nouns a group of other nouns which, following the example of the Arabic grammarians, we call *nomina* (adjectiva) *relativa*, i. e. nouns of relation. Thus أَرْضِيّ belonging to the earth (أَرْض), earthly; شَامِيّ belonging to شَام (i. e. Syria), a Syrian. The feminine termination is dropped when this ending is added, as مَكِّيّ (from مَكَّة) an inhabitant of Mecca; occasionally we meet with certain changes in the vowels of a word, e. g. مَدَنِيّ an inhabitant of Medina, from ٱلْمَدِينَة Medina; قُرَشِيّ a Ḳoreishite, one of the tribe قُرَيْش.

b. By the addition of the feminine ending to nouns of relation there are formed feminines, as شَامِيَّة a Syrian woman, but more frequently abstract nouns; as إِلَاهِيَّة divinity from إِلَاهِيّ divine, (from إِلَاه God); جَاهِلِيَّة heathenism from جَاهِلِيّ heathenish, (from جَاهِل ignorant).

NOTE. It is usual to indicate the nomina relativa also by paradigms from فَعَلَ; thus we say that أَرْضَى is a form فَعْلَى, جَاهِلِيَّةٌ a form فَاعِلِيَّةٌ.

Diminutives from triliteral nouns take the form 66. فُعَيْلٌ, as عُبَيْدٌ a little slave, servulus, from عَبْدٌ slave. From quadriliteral nouns the form is فُعَيْلِلٌ, as عُقَيْرِبٌ a little scorpion, from عَقْرَبٌ (so صُوَيْحِبٌ diminutive from صَاحِبٌ companion). From quadriliteral nouns with a long vowel between the third and fourth radicals the corresponding form is فُعَيْلِيلٌ, as صُنَيْدِيقٌ diminutive from صُنْدُوقٌ a box. Diminutives are not unfrequently derived also from proper names, as عُبَيْدُ ٱللّٰهِ 'ubaidullāhi alongside of عَبْدُ ٱللّٰهِ 'abdullāhi (Abdallah).

The formation of nouns from stems mediae gemi- 67. natae and from those with a hamza or the semi-vowels presents many irregularities, for a general idea of which we must refer to the inflection of the corresponding verbal stems. In addition to what is there given the following particulars deserve attention.

For the formation of deverbal nouns from stems mediae geminatae (see § 34 ff.) the following points may be noted:

The second and third radicals are of course con- *a.*

tracted when the second is without a vowel of its own, as فَرّ from فَرَرْ.

b. If the first radical has *a*, and the second *i* or *a*, contraction takes place in the participles and infinitives, e. g. part. act. VII of فَرَّ: مُنْفَرّ contracted from مُنْفَرِر; pass. also مُنْفَرّ from مُنْفَرَر. There is no contraction, however, with nouns of the form فَعَلٌ, as دَبَبٌ inf. to be hairy.

c. According to the rule given in § 35 *b*, from مَفْرَر we get مَفَرّ; from مُفْرِر : مُفِرّ.

d. The act. participle of I is فَارّ from فَارِر cf. § 8.

e. Contraction does not take place when a long vowel stands between the last two radicals e. g. فِرَار, مَفْرُور, فَرِير.

68. The orthographical rules which apply to the inflection of the verba hamzata (§§ 37 ff.) hold good for the formation of nouns, e. g. سُؤْل something asked for; سُؤَال a question, from سَأَل to ask; the part. act. I of أَثَر, to make an impression, is آثِر for أَأْثِر; مِثْبَرَة NF. nomen instrumenti مِفْعَلَة from أَثَر &c.

69. The primae و stems, which according to § 40 lose
a. their first radical in the impf., lose it also, as a rule,

in the nomen verbi; as compensation the latter receives the feminine termination (§ 73), as from وَعَدَ to promise nomen verbi عِدَةٌ; from وَدَعَ to allow: دَعَةٌ.

w after the vowel *i* (ِو) coalesces with the latter b. to form *ī*, as inf. IV of وَقَعَ fall: إِيقَاعٌ for إِوْقَاعٌ; مِيلَادٌ time of one's birth NF. مِفْعَالٌ, for مَوْلَادٌ from وَلَدَ.

ِو passes into *ū* (§ 40 c), e. g. part. IV of يَقِظَ c. to be awake: مُوقِظٌ for مُيْقِظٌ.

In the infs. of the IV. and X. stems from *stems* 70. *med.* و *and* ى the middle radical disappears; the a. feminine termination is added as compensation, e. g. إِقْوَالٌ for إِقَالَةٌ.

In the act. part. of stem I the *w* of verbs med. و b. becomes *y* and ِي (*yi*) is changed into *'i* (5); as قَائِلٌ for قَاوِلٌ, سَائِرٌ for سَايِرٌ (for Medda see § 7).

A characteristic formation from these stems is فَيْعِلٌ; c. thus from the stem ساد med. و we get سَيِّدٌ master, lord; from the stem طاب med. ى, طَيِّبٌ good.

Nouns formed on the model of فَعْلٌ contain diph- d. thongs (§ 2 a), as سَيْرٌ, قَوْلٌ.

The place of the second radical (see § 42) is taken e. by a long *ā* in the act. participles of stems VII. and

VIII. and in the pass. part. of stems IV., VII., VIII. and X.; e. g. part. pass. IV. مُقَامٌ, part. act. or pass. VII. مُنْقَامٌ (from a hypothetical active مُنْقَوِمٌ pass. مُنْقَوَمٌ). Also in numerous nominal forms, as دَارٌ (from a hypothetical دَوَرٌ) house, from دَارَ med. و; NF. مَفْعَلٌ from قَالَ is مَقَالٌ, from a hypothetical مَقْوَلٌ.

f. The place of the second radical (see § 43) is taken by a long *ī* in nouns of the type of فِعْلٌ and فِعْلَةٌ from med. و and ى e. g. لِينٌ from لَانَ med. ى to be gentle; مِيتَةٌ (§ 64 c) for مِوْتَةٌ mode of death from med. و; in the form فُعْلٌ from med. ى, e. g. بِيضٌ for بُيْضٌ white (plur.); مَفْعِلٌ in the forms from med. ى, e. g. مَسِيرٌ, walk for مَسْيِرٌ; in the part. act. of the IV. and X. stems from verbs mediae و and ى, e. g. مُقِيمٌ, مُسْتَسِيرٌ; in the part. pass. I from med. ى, e. g. مَبِيعٌ from بَاعَ, to sell (mediae ى) for مَبْيُوعٌ.

g. The place of the second radical is taken by long *ū* in nouns of the type of فُعْلٌ from med. و, as نُورٌ light from نَارٌ; *ū* may also arise by contraction from *wū* in the pass. part. of the I stem of verbs med. و, as مَقُولٌ for مَقْوُولٌ.

71. In the case of nouns derived from verbs ultimae
a.

71. NOUNS FROM STEMS ULTIMAE و AND ى.

و and ى those forms in which the second radical is vowelless are treated like forms from strong stems, as غَزْوٌ, رَمْىٌ inf.

If the second radical has ă, there results (cf. § 46 a) *b.* at the end of words a long *ā* (from hypothetical *awu, ayu*) which is written ـَا or ـَى (acc. as last rad. is و or ى), e. g. اَلْعَصَا the stick, for اَلْعَصَوُ; اَلْمَرْعَى the pasture, from رَعَى to feed, for a hypothetical اَنْفَعَل NF. اَتْخَى for أَتْخَىُ, elative of سَخِىٌ generous, liberal (§ 63 *b*). The same applies to all the pass. participles of the derived stems. With the nunation, these forms appear as مُرْمًى, مَرْعًى, عَصًا (ptc. pass. IV) in which the original long final vowel, now standing in a syllable closed by the *n* of the nunation, must be pronounced short (§ 8): *'aṣan, mar'an, murman*. Long *ā* appears before the feminine termination (cf. § 70 *e*) as, غَدَاةٌ morning for غَدَوَةٌ; وَفَاةٌ death for وَفَيَةٌ.

If the second radical has short ĭ, from *iyu* arises *c.* a long *ī* (cf. § 47 *a*), e. g. اَلرَّامِى part. act. I in place of a hypothetical اَلرَّامِىُ; and so in the act. participles of the derived forms. If the nunation is added, the result is رَامٍ, *rāmin* &c., in which the ى is dropped even in the written form of the word. *ŭyu* is changed to *iyu*, and consequently with the nunation it likewise becomes *in*; e. g. inf. V. تَرَمّ ; اَلتَّرَمِّي for اَلتَّرَمُّيُ for تَرَمُّيٌ. In the act. part. of stem I from verbs ult. و, *iwun* is changed to *iyun*, and consequently with the nunation further to *in*, e. g. اَلْغَازِو, اَلْغَازِي for اَلْغَازِيُ; with the nunation غَازٍ. Before *ă* and *ā* (cf. § 47 *d*), on the other hand, the third radical retains its consonantal value; thus the inf. of stem II, according to the form most in use with verbs med. و and ي viz. تَفْعِلَة (§ 61), is: تَغْزِيَة ,تَرْمِيَة.

d. After *ā*, *yu* and *wu* become *'u*; *yun*, and *wun* become *'un*, in each case with the hamza, e. g. اَلسَّرَآءُ for اَلسَّرَاوُ with the nunation سَرَآءٌ inf. I of سَرُوَ to be noble; اَلْإِرْمَآءُ for اَلْإِرْمَايُ, with the nunation إِرْمَآءٌ inf. IV for إِرْمَايٌ.

e. If the second radical has a long *ū*, the forms from verbs ultimae و are formed regularly; thus the pass. part. I of غَزَا is مَغْزُو (for مَغْزُوو) *maḡzūwun*. From verbs ultimae ي, on the other hand, *ūyun* is changed to *īyun*, e. g. مَرْمِي (from مَرْمُوي) *marmīyun*, so from مَضَى go away inf. مُضِي for مُضُوي NF. فُعُول.

f. If the second radical has a long *ī*, the forms from

verbs ultimae ى are formed regularly, e. g. NF. فَعِيلٌ from وَلِيَ: وَلِيٌّ saint (for وَلِيْيٌ) walīyun. From verbs ultimae و, on the other hand, iwun is changed into iyun, as عَلِيٌّ 'alīyun high from عَلِيوٌ.

b. *The Gender of Nouns.*

72. Arabic has two genders, a masculine and a feminine. A number of words are sometimes masculine sometimes feminine, in other words are of the common gender. Words which denote female beings, collectives, countries, cities, winds, parts of the body occurring in pairs, and others, are in themselves feminine without requiring the feminine termination. The gender of such words is in each case noted in the dictionaries.

73. *a.* As an outward and visible sign of the feminine we find most frequently the ending ةٌ ـَ atun (or ةٌ ـَ atu § 79), e. g. قَاتِلَةٌ (NF. فَاعِلَةٌ), fem. of قَاتِلٌ killing; مَلِكَةٌ (NF. فَعَلَةٌ) queen, from مَلِكٌ; رَاضِيَةٌ fem. of masc. رَاضٍ (§ 71 c) content, فَتَاةٌ (NF. فَعَلَةٌ) maid, from فَتًى (§§ 71 b and 2 d) youth. Many substantives are found only with the feminine ending, as جَنَّةٌ an orchard.

NOTE. As a rarity, the feminine ending is found, particularly in the Ḳur'ān, written with ت, e. g. نِعْمَتُ ٱللّٰهِ the grace of God (for نِعْمَةٌ).

b. A number of masc. nouns are found with the feminine ending, as خَلِيفَةٌ Caliph, طَلْحَةُ Ṭalḥa (proper name of a man, see p. 8, note 2). On the other hand, there are nouns which, as being essentially feminine, do not require the feminine termination, as عَاقِرُ barren (referring to a woman).

c. The feminine ending ةٌ‏َ is occasionally appended to common or class nouns in order to indicate a single individual (nomen unitatis), as ذَهَبَةٌ a gold piece, from ذَهَبٌ gold; حَمَامَةٌ a dove, from حَمَامٌ doves (collective). The termination ةٌ‏َ is also used for the formation of the so-called nomina vicis, i. e. nouns that express the doing of an action *once*, as تَعْدَةٌ a single sitting down, from تَعَدَ to sit down.

d. The feminine termination, again, serves to form substantives from adjectives, as سَاقِيَةٌ conduit-pipe, water-channel, from the part. I of سَقَى to water. Connected probably with this is the feminine ending which forms intensives, as عَلَّامَةٌ a very learned person, from the adjective عَلَّامٌ § 63 *a*.

e. Collective nouns are also formed by means of the feminine termination, e. g. from رَكَّاضٌ a courier, coll. رَكَّاضَةٌ; صُوفِيٌّ (§ 65 *a*) Ṣūfī (mystic), coll. صُوفِيَّةٌ.

74. Other feminine terminations are:

a. The termination ى‎ــَ‎; it goes to form feminines of the type فُعْلَى‎, e. g. سَكْرَى‎ fem. of سَكْرَانُ‎, drunk, (§ 58 b); feminines of the nominal form (NF.) فُعْلَى‎ from elatives (§ 63 b), e. g. صُغْرَى‎ fem. of أَصْغَرُ‎ smaller, أُولَى‎ from أَوَّلُ‎ the first, and substantives like دُنْيَا‎ world (§ 2 note), which is properly a feminine to the elative أَدْنَى‎, that which is nearer at hand; also feminines of the NF. فِعْلَى‎, e. g. from أَحَدُ‎ one, fem. إِحْدَى‎; subst. ذِكْرَى‎ remembrance.

b. The ending ـَاءُ‎; it goes to form, more especially, adjectives of the NF. فَعْلَاءُ‎ from أَفْعَلُ‎ (§ 62 c), e. g. صَفْرَاءُ‎ fem. yellow; عَوْرَاءُ‎ fem. one-eyed, but also substantives, as صَحْرَاءُ‎ desert.

c. Inflection of the Noun.

75. Arabic has three *numbers*: singular, dual and plural. Of the last, there are two different kinds; the one, the ordinary plural, properly so called, also known as the pluralis sanus or the outer plural, which originally denoted rather a number of separate persons and things; the other, the collective plural, also called the inner or broken plural (see §§ 86 ff.), which denotes rather a continuous mass, in which the individual member is not distinguished. At present we shall deal only with the first-named. Arabic distinguishes three cases: Nominative, Genitive, and Accusative.

76. The terminations of the dual and the pluralis
a. sanus are as follows:

Dual nominative	ــَانِ	(cf. § 33)
„ genitive and accusative	ــَيْنِ	(cf. םִ‍ַ)
Plural mascul. nominative	ــُونَ	(cf. § 33)
„ „ gen.-accus.	ــِينَ	(cf. םִ‍ַ)
„ femin. nominative	ــَاتٌ	(cf. ות)
„ „ gen.-accus.	ــَاتٍ	

Before these terminations the flectional endings of the sing. are dropped; the ة of the feminine ending is changed to ت before the dual termination, (as it is before the pronominal suffixes appended to the singular), e. g. جَارِيَةٌ, dual جَارِيَتَانِ.

b. By the addition of the terminations exhibited above is formed the plural of many adjectives, in particular, and also of a number of substantives. In the formation of the plural we find substantives with the feminine ending taking the sign of the masculine plural (as سَنَةٌ year, plur. سِنُونَ); much more frequently, however, substantives without the sign of the feminine in the singular are found forming their plural by means of the feminine termination, e. g. حَالٌ condition, plur. حَالَاتٌ, سَمَاءٌ heaven, plur. سَمَاوَاتٌ (with the original wāw restored § 71 *d*), also written سَمَوَاتٌ.

As regards the case inflection of the singular, it **77** is necessary to distinguish between the so-called nomina triptota or triptotes, *i. e.* nouns which are inflected for all three cases, and the so-called nomina diptota or diptotes, *i. e.* nouns which cannot be thus fully inflected. The latter never receive the nunation, and unless they are <u>determined by the article or by a following genitive, they are</u> inflected for only two cases.

The following are the case-endings of the triptote *a*. noun: Nom. sing. ـٌ *un*, Gen. sing. ـٍ *in*, Acc. sing. ـًا *an*. With the feminine termination ـً only is written instead of ـًا as رَجُلٌ, but مَدِينَةٍ; so فَتًى and عَصًا (cf. § 3 *b*).

The case-endings of the diptote noun are: Nom. *b*. sing. ـُ *u*, Gen. and Accus. Sing. ـَ *a*.

In the dictionary the triptotes are distinguished from the diptotes by being always written with the

5*

nunation, as رَجُلٌ a man, while the latter are always without it, as أَسْوَدُ black.

78. Whole classes of nouns are always diptote. Such are

a.   1) all proper names that are either feminine or have the feminine termination, as مَيَّةٌ, زَيْنَبُ as names of women; مَسْلَمَةُ as name of a man. To these must be added the majority of such proper names as are of foreign origin, e. g. إِبْرَاهِيمُ Abraham, يُوسُفُ Joseph, مُوسَى Moses (but monosyllables like نُوحٌ Noah are mostly triptote).

b.   2) Many so-called broken plurals; cf. § 88 Nos. 18, 19, 20; § 89 Nos. 23 24, 25, 27, 29;

c.   3) adjectives of the form أَفْعَلُ (§ 62 *c*; § 63 *b*);

d.   4) adjectives of the form فَعْلَانُ (§ 58 *b*), which form their fem. like فَعْلَى, e. g. غَضْبَانُ angry, fem. غَضْبَى.

e.   5) Feminines formed by the terminations ـَى or ـَاءُ (§ 74). Cf. also the broken plurals referred to under *b*, §§ 88,19 and 89,29.

79. The inflection of the singular of all nouns and of the plural of feminines varies according as a noun is *determined* or *undetermined*.

a.   All proper names are in themselves determined as مُحَمَّدٌ *muḥammadun* Muhammed; أَحْمَدُ *aḥmadu*

Ahmed; such proper names are treated either as triptotes or as diptotes according as their form and the custom of the language may determine; many of them always take the article, as اَلْحَارِث.

Common or class nouns are determined:

1) by the article; as فَرَسٌ a horse, اَلْفَرَسُ the horse. b.

2) by the addition of a following genitive, which c. may be either a noun or a pronominal suffix, whereby the nomen regens is put in the *construct state*; as فَرَسُ الرَّجُلِ the horse of the man, فَرَسُهُ his horse.

The case-endings of a noun determined (1) by the prefixing of the article, or (2) by a genitive following —and the same applies to proper names with the article—are distinguished as follows from those of the undetermined noun:

Singular nom. ُ , Gen. ِ , Acc. َ .

Plural fem. nom. ُ , Gen.-Acc. ِ

i. e. the nunation is always dropped. These endings are assumed not merely by all triptotes, but also by the diptotes, when determined by the article or a genitive following: e. g. Nom. أَسْوَدُ, Gen.-Acc. أَسْوَدَ; but Nom. اَلْأَسْوَدُ, Gen. اَلْأَسْوَدِ, Acc. اَلْأَسْوَدَ.

Before a following genitive (which acc. to § 79 c **80.** may be either a noun or a pronominal suffix) the terminations ن of the dual and ن of the plural are dropped, thus:

Dual Nom. of عَبْدٌ: عَبْدَانِ, but عَبْدَا ٱلْوَزِيرِ the two slaves of the Vizier.

Dual Gen.-Acc. عَبْدَيْنِ, but ضَرَبْتُ عَبْدَيْ عُمَرَ I have beaten the two slaves of Omar (before a connective Alif thus: عَبْدَيِ ٱلْوَزِيرِ, cf. § 6 e).

Plural Nom. of قَصَّابٌ butcher, executioner قَصَّابُونَ, but قَصَّابُو ٱلْمَلِكِ the executioners of the king.

Plural Gen.-Acc. قَصَّابِينَ, but رَأَيْتُ قَصَّابِي ٱلْمَلِكِ I have seen the executioners of the king.

For the inflection of the noun see paradigms XX and XXI, where will be found the forms of the masculine triptote قَصَّابٌ an executioner, the masculine diptote آخَرُ another, the feminine triptote سَاعَةٌ hour, and the feminine diptote مَيَّةُ Mayya (name of a woman).

81.

a. In the case of nouns derived from stems ultimae و and ى when the second radical has a short vowel the nunation, acc. to § 71 b c, is taken by this vowel of the second radical.

b. Nouns ending in *an* or *ā* are unchangeable for all three cases; those in *in* or *ī*, on the other hand, take the *an* of the nunation, as well as the simple *a* (§ 47 d) as رَامِيًا, اَلرَّامِىَ.

82. THE ADDITION OF THE PRONOMINAL SUFFIXES.

Before the dual terminations (cf. § 46 d) the last *c.* radical is treated as a strong letter, as مَرْعَيَانِ, عَصَوَانِ, رَامِيَانِ.

In the plural the last radical is dropped before *d.* the terminations *ūna* and *īna*, which, when joined to an *a* of the second radical, produce diphthongs (§ 46c); thus from مَرْمًى: مَرْمَوْنَ, مَرْمَيْنَ; if the second radical has *i*, the terminations are added immediately to the former (§ 47c), as رَامُونَ, رَامِينَ.

For the inflection of these nouns see paradigm No. XXII, where will be found the forms of the triptote قَاضٍ judge, the triptote مُصْطَفًى (ult. ى) chosen one (often as a proper name), the triptote عَصًا (ult. و) a stick, the diptote ذِكْرَى remembrance, and the diptote دُنْيَا world (vgl. § 74 a).

For the forms of the pronominal suffixes see 82. § 12 b—d.

Before the pronom. suffix of the 1. pers. sing. the *a.* short case-endings of the construct state are dropped, as قَصَّابِي. The said suffix after a final *ā, ī* or *ai* becomes َى (*ya*), as with the nom. dual قَصَّابَايَ, with ـَى: فَتَى (§ 2 d; 81 a); with the gen.-acc. plur. قَصَّابِيَّ; with ـِى: قَاضِيَّ (§ 81 a); with gen.-acc. dual قَصَّابَيَّ.

NOTE. In the case of words which end in ىً, the suffix may either be attached in the usual way, e. g. from بُنَىً "sonny", بُنَىِّ, or appended to the shortened form ىً——, e. g. بُنَىَّ from بُنَىً and ى.

b. The final *ū* of the construct state of the plural masc. is changed to *ī* before the appended ى (cf. § 71 *e*), thus قَصَّابُو becomes قَصَّابِى, and then with the suffix of the 1. pers. sing. قَصَّابِىَّ (no longer to be distinguished from the genit. and accus. plural). The same applies to the ending *au* from stems ult. ى (see parad. XXII), e. g. مُصْطَفَوْ becomes مُصْطَفَىْ, with the suffix مُصْطَفَىَّ (also identical with the genitive-accusative form).

For the union of the noun with the suffixes see paradigm XXIII. For the change before suff. of final ة into ت see § 76 *a*.

83. In the pluralis sanus of substantives of a masc. or fem. nominal form with one short vowel (that is, of any of the following types فَعْلٌ, فِعْلٌ, فُعْلٌ and فَعْلَةٌ, فِعْلَةٌ, فُعْلَةٌ) the second radical frequently receives a *complementary vowel* which is either identical with that of the first radical or is short *ă*. Thus أَرْضٌ earth, plur. أَرَضُونَ, more rarely أَرْضُونَ, and أَرَضَاتٌ, more rarely أَرْضَاتٌ; ظُلْمَةٌ darkness, plur. ظُلُمَاتٌ alongside

of ظُلَمَاتٌ and ظُلُمَاتٌ. This is a favourite method in the case of the plural of the form فَعْلَةٌ, as طَعْنَةٌ (§ 73c) a single thrust or blow; plur. طَعَنَاتٌ several thrusts or blows.

84. Before اِبْن a son, a proper name loses its nunation in the case mentioned § 6f 2, and اِبْن is itself written without the prosthetic ا, e. g. مُسْلِمُ بْنُ ٱلْوَلِيدِ *muslimu-bnu-lwalīdi* Muslim, the son of al-Walīd. زَيْدٌ ٱبْنُ بِشْرٍ *zaiduni-bnu bischrin* (§ 6e) means, on the other hand, Zaid is the son of Bishr (nominal sentence).

85. After يَا the particle of address, the simple noun follows in the nominative without the nunation, as يَا مُحَمَّدُ Muhammed, يَا رَجُلُ Oh M.! يَا رَجُلُ Oh man! (by which a definite person is hailed). But should anything of the nature of a complement (a genitive, for instance) be added to the noun in the vocative, the name of the person addressed must be put in the accusative, as يَا عَبْدَ ٱللّٰهِ: عَبْدُ ٱللّٰهِ o Abdallah! (Oh servant of God!); يَا بَنِى كِنْدَةَ Oh Banu Kinda! *i. e.* members of the tribe of Kinda (here بَنِى cf. § 80 and 90b is the constr. state of بَنِينَ). If an Object follows, the noun stands in the accus. with the nunation, as

يَا رَاكِبًا ٱلْحَمْرَاءِ Oh thou that ridest the red mare! — The particle أَيُّهَا (before which we may also have يَا) is always followed by a nominative with the article, as يَا أَيُّهَا ٱلنَّاسُ Oh ye people!

NOTE. After يَا, which serves as the expression of pain and sorrow, a long $\bar{a}$ is appended to the noun; in pause اه, as يَا أَمَاهْ Oh mother!

86. There are, in Arabic, a mass of words which, though singular in form, have a *collective* signification. The following varieties may be singled out under this head:

a. Simple collectives (masc. gend.) such as قَوْم, which denotes not merely 'a people' collectively, but also 'people' as individuals; عَسْكَر an army and also the individual soldiers thereof. From such words broken plurals may be formed.

b. Names of the inhabitants of a country, as ٱلْيَهُودُ the Jews, often coinciding with the name of the country itself, as ٱلْهِنْدُ the Hindus; a single Jew or Hindu is called يَهُودِيّ, هِنْدِيّ § 65 a.

c. Class names (masc. gend.) from which are formed nomina unitatis (§ 73 c) as حَمَام doves.

d. So-called quasi-plurals (masc. gend.), from which no nomen unitatis is formed, as رَكْب a company of horsemen (a single one رَاكِبٌ)؛ خَدَمٌ the domestics (one of which is خَادِمٌ)؛ حَمِيرٌ a number of asses (one ass حِمَارٌ)؛ عَبِيدٌ slaves (from عَبْدٌ).

87. The so-called *broken plurals* (plurales fracti in the language of the native grammarians—by German scholars by preference called 'inner plurals' because due to changes in the body of the word) are also strictly speaking nothing more than collectives. Hence they are treated in Arabic as singular nouns of the feminine gender and construed accordingly. Thus أَبْوَابٌ مُتَفَرِّقَةٌ different gates, where أَبْوَابٌ is the broken plural of بَابٌ (on the model of أَفْعَالٌ), and the participle act. V. of فَرَقَ is put in the fem. sing.—These broken plurals, further, take the same inflection as the singulars, discussed in § 77 ff.

b. As a rule the broken plurals are given in the dictionaries alongside of the singular of their respective nouns; when this is not so, it is to be presumed that the word either has no plural or takes a pluralis sanus. Sometimes we find from one and the same word more than one plural; in such a case, not unfrequently, a word varies its plural as its meaning varies. Certain of the broken plurals are, as a rule, confined to certain specified singulars.

88. THE BROKEN PLURALS.

88. From nouns regarded as containing three consonants the following broken plurals may be formed:

1. فُعْل from أَفْعَل (§ 62 c) and its fem. فَعْلَاء (§ 74 b), as حُمْر from أَحْمَر red; سُود (cf. § 70 g) from أَسْوَد black; بِيض (for بُيْض cf. § 70 f) from أَبْيَض white.

2. فُعُل from various singulars, as كُتُب from كِتَاب book.

3. فِعَل from sing. فِعْلَة, as قِطَع from قِطْعَة piece.

4. فُعَل mostly from sing. فُعْلَة, as عُلَب from عُلْبَة box; أُمَم from أُمَّة people; occasionally from فَعْلَة, as قُرًى (for قُرَي) acc. to § 71 b) from قَرْيَة place.

5. فِعْلَة, as إِخْوَة from أَخ brother.

6. فَعَلَة esp. from sing. فَاعِل, as كَمَلَة from كَامِل perfect; but also from فَعِيل § 70 c, as سَادَة (for سَيَدَة) from سَيِّد lord.

7. فِعَلَة (rare) as قِرَدَة from قِرْد monkey,

8. فُعَلَة from فَاعِل ult. ى, as قُضَاة (for قُضَيَة) § 71 b) from قَاضٍ judge.

9. فِعَال very common, from various singulars, as قِدَاح from قِدْح arrow.

88. THE BROKEN PLURALS.

10. فُعُول very common, also from various singulars, as جُنُود from جُنْد band of soldiers; بُكِيّ (for بُكُوي see § 71 e) and then (with change of u to i) بِكِيّ from بَاكٍ weeping.

11. فِعَالَة (rare) as حِجَارَة from حَجَر stone.

12. فُعُولَة (rare) as عُمُومَة from عَمّ uncle.

13. فُعَّل from فَاعِل, as بُهَّل from بَاهِل an unbranded she-camel.

14. فُعَّال from فَاعِل, as كُتَّاب from كَاتِب scribe.

15. أَفْعُل from various singulars, as أَرْجُل from رِجْل foot.

16. أَفْعِلَة from various singulars, as أَرْغِفَة from رَغِيف a cake, أَحِبَّة (§ 67 c) from حَبِيب beloved; أَئِمَّة from إِمَام president; آلِهَة from إِلَاه God.

17. أَفْعَال very common, from various singulars, as أَمْطَار from مَطَر rain; أَشْيَاء (always without the nunation) from شَيء thing.

18. أَفْعِلَاء esp. from فَعِيل, as أَقْرِبَاء from قَرِيب relative; أَغْنِيَاء from غَنِيّ rich.

19. فَعْلَى (rare), as جَرْحَى from جَرِيح wounded.

20. فُعَلَاءُ, as شُعَرَاءُ from شَاعِرٌ poet.

21. فِعْلَانٌ, as فِتْيَانٌ from فَتًى youth; جِيرَانٌ (for جِوْرَانٌ cf. § 69 b) from جَارٌ neighbour.

22. فُعْلَانٌ, as بُلْدَانٌ from بَلَدٌ district; فُرْسَانٌ from فَارِسٌ rider; سُودَانٌ negroes from أَسْوَدُ black.

NOTE. Forms 5 and 15—17 are used, as a rule, only of a number of objects not exceeding ten (hence called pluralia paucitatis).

89. From nouns with more than three radical consonants (cf. § 56 d ff.) are formed plurals in which the first consonant takes *ă*, the second *ā* and the third *i*. Such plurals are diptotes with the exception of all those derived from stems ult. ي (or with an additional ي in the sing. § 74 a) which take the nunation *in* in the nominative and genitive, but not in the accusative which ends in ـِيَ. The forms of the singular of Nos. 24 (cf. also صُوَيْجِبٌ § 66) and 25 are regarded as quadriliterals. No. 29 ends in long *ā* and is diptote. The following are the principal varieties:

23. فَعَالِلُ as جَنَادِبُ from جُنْدَبٌ (NF. فُعْلَلٌ) locust. This form is also found from nouns that are only in a special sense quadriliterals, inasmuch as they are really triliterals with the addition of a

formative consonant; examples of this group are:
a) أَفَاعِلُ, as أَنَامِلُ from أَنْمُلَةٌ (NF. أَنْعُلَةٌ) fingertip;
also from elatives used as substantives, such as
تَفَاعِلُ (b) ; كَبِيرٌ elat. of أَكَابِرُ the great ones from
مَفَاعِلُ (c) experience; (تَفْعِلَةٌ .NF) تَجْرِبَةٌ from تَجَارِبُ as
as مَزَابِلُ from مَزْبَلَةٌ (NF. مَفْعَلَةٌ) dung-heap; مَعَايِشُ
(with ي, not with ئ) from مَعِيشَةٌ (NF. مَفْعِلَةٌ) livelihood;
مَعَانٍ (acc. مَعَانِيَ) of مَعْنًى (NF. مَفْعَلٌ) idea.

24. فَوَاعِلُ especially from فَاعِلَةٌ and فَاعِلٌ (used
as a substantive), as صَوَاعِقُ from صَاعِقَةٌ thunder-clap;
§ 67 b) خَوَاصُّ (for) خَوَاصِصُ from فَارِسٌ rider; فَوَارِسُ
from خَاصٌّ person of distinction; جَوَارٍ (acc.
جَوَارِيَ) from جَارِيَةٌ a female slave.

25. فَعَائِلُ from such nominal forms with a long
vowel after the second radical as have a feminine
form or signification, as a) جَنَائِزُ from جِنَازَةٌ funeral
obsequies; b) عَجَائِبُ from عَجِيبَةٌ miracle; c) عَرَائِسُ
from عَرُوسٌ bride.

26. فَعَلَى) فَعَالٍ as فَتَاوٍ from فَتْوَى (N. F. فَعْلَى) decision.
27. فَعَالِيلُ from quadriliteral nouns with a long
vowel before the last consonant, as عَنَاقِيدُ from عُنْقُودٌ

(N. F. يُعْلُولٌ) bunch of fruit; this form is also found with nouns derived from triliteral stems, of which the following are specimens: a) أَحَادِيثُ as أَفَاعِيلُ from أُحْدُوثَةٌ (NF. أُفْعُولَةٌ) story; b) تَفَاعِيلُ as تَصَارِيفُ from تَصْرِيفٌ (infinitive تَفْعِيلٌ used as a noun) turn; c) مَقَاعِيلُ as مَقَادِيرُ from مَقْدُورٌ (participle مَفْعُولٌ used as a noun) fate; but also جَوَاعِيلُ (cf. No. 24) as جَوَاسِيسُ from جَاسُوسٌ (NF. فَاعُولٌ) spy.

28. فَعَالِلَةٌ, from quadriliteral nouns denoting living beings, as جَبَابِرَةٌ from جَبَّارٌ (NF. فَعَّالٌ) a mighty man; أَسَاقِفَةٌ from أُسْقُفٌ bishop; تَلَامِذَةٌ from تِلْمِيذٌ pupil; بَغَادِدَةٌ from بَغْدَادِيٌّ a native of Bagdad.

29. فَعَالَى, as صَحَارَى from صَحْرَاءُ desert; هَدَايَا (for هَدَايِىُ § 2 d note b) from هَدِيَّةٌ (NF. فَعِيلَةٌ) from ult. ى) present.

90. The following nouns (arranged in alphabetical order) are more or less irregular in their mode of inflection:

a. أَبٌ father, أَخٌ brother and حَمٌ father-in-law take the following forms in the construct state and before suffixes beginning with a consonant:

90. IRREGULAR NOUNS.

Nominative   حَمُو، أَخُو، أَبُو
Genitive   حَمِى، أَخِى، أَبِى
Accusative   حَمَا، أَخَا، أَبَا

The Dual of أَبٌ is أَبَوَانِ (i. e. the two parents), the plur. آبَاءٌ (§ 88 No. 17). The vocative singular with suff. of the 1. pers. sing. of أَبٌ is يَا أَبَتِ, يَا أَبَنِى, يَا أَبَتَ; from أَخٌ: أَخِى; with suffix of the 2. pers. masc. sing. أَخُوكَ, أَبُوكَ.

b. اِبْنٌ son; plur. sanus has nom. بَنُونَ (construct بَنُو), gen.-acc. بَنِينَ (st. constr. بَنِى); broken plur. أَبْنَاءٌ (§ 88,17).

c. أَخٌ brother, see a; broken plur. إِخْوَانٌ, إِخْوَةٌ (§ 88,5. 21).

d. أُخْتٌ sister; plur. أَخَوَاتٌ.

e. اِمْرُءٌ or اِمْرُؤٌ (also مَرْءٌ) man; gen. اِمْرِئٍ; acc. اِمْرَأً.

f. اِمْرَأَةٌ woman; plur. from another root نِسَاءٌ, نِسْوَةٌ or نِسْوَانٌ (§ 88,9. 5. 21).

g. أُمٌّ mother; plur. أُمَّاتٌ or أُمَّهَاتٌ.

h. إِنْسَانٌ man, human being; plur. أُنَاسٌ, collective نَاسٌ.

Socin, Arabic Grammar.²

90. IRREGULAR NOUNS.

i. بِنْتٌ daughter, frequently also اِبْنَةٌ (with connective Alif); plur. بَنَاتٌ.

k. دِينَارٌ dinar, gold-piece; broken plur. irregular, دَنَانِيرُ.

l. ذُو (only in the st. constr.) possessor of ....; gen. ذِي, acc. ذَا; fem. ذَاتُ; dual nom. ذَوَا; plur. nom. ذَوُو (gen.-acc. ذَوِي) fem. ذَوَاتُ; for the plural أُولُو (*ŭlū*), gen.-acc. أُولِي is used.

m. سَنَةٌ year; plur. nom. سِنُونَ (or سُنُونَ); gen.-acc. سِنِينَ.

n. عَمْرٌو *'amrun*, 'Amr, proper name of a man. A و is added to the written form of this word in the nom. and gen. (عَمْرٍو) to distinguish it from عُمَرُ *'umaru* (a diptote). Acc. عَمْرًا; followed by بْنَ it is written عَمْرو and pronounced *'amra-bna*.

o. فَمٌ or فُوهٌ mouth; st. constr. usually nom. فُو, gen. فِي, acc. فَا; broken plur. (§ 88,17) أَفْوَاهٌ.

p. لَيْلٌ night; broken plur. (from the root ليلى) لَيَالٍ (§ 89,23).

q. مَاءٌ water; broken plur. مِيَاهٌ or أَمْوَاهٌ (§ 88,9.17).

r. يَدْ hand; broken plur. (§ 88,15) أَيْدٍ from أَيْدٍى (cf. § 71 c).

s. يَوْمْ day; broken plur. أَيَّامْ from أَيْوَامْ (§ 88,17).

Chapter IV. The Numerals. (§§ 91—93.)

91. The cardinal numbers have the following forms:

	Masc.	Fem.	
1	وَاحِدْ	وَاحِدَةْ	inflected
	أَحَدْ	إِحْدَى	,,
2	اِثْنَانِ	اِثْنَتَانِ	(inflected as a dual)
3	ثَلَاثْ (ثَلْثْ)	ثَلَاثَةْ (ثَلْثَةْ)	inflected
4	أَرْبَعْ	أَرْبَعَةْ	,,
5	خَمْسْ	خَمْسَةْ	,,
6	سِتْ	سِتَّةْ	,,
7	سَبْعْ	سَبْعَةْ	,,
8	ثَمَانٍ (see p. 27*)	ثَمَانِيَةْ	,,
9	تِسْعْ	تِسْعَةْ	,,
10	عَشْرْ	عَشَرَةْ	,,
11	أَحَدَ عَشَرَ	إِحْدَى عَشْرَةَ	indeclinable

91. THE CARDINAL NUMBERS.

	Masc.	Fem.	
12	اِثْنَا عَشَرَ	اِثْنَتَا عَشْرَةَ	gen.-acc. اِثْنَىْ عـ, اِثْنَتَىْ عـ
13	ثَلَاثَةَ عَشَرَ	ثَلَاثَ عَشْرَةَ	indeclinable
14	أَرْبَعَةَ عَشَرَ	أَرْبَعَ عَشْرَةَ	,,
15	خَمْسَةَ عَشَرَ	خَمْسَ عَشْرَةَ	,,
16	سِتَّةَ عَشَرَ	سِتَّ عَشْرَةَ	,,
17	سَبْعَةَ عَشَرَ	سَبْعَ عَشْرَةَ	,,
18	ثَمَانِيَةَ عَشَرَ	ثَمَانِيَ عَشْرَةَ	,,
19	تِسْعَةَ عَشَرَ	تِسْعَ عَشْرَةَ	,,

20 عِشْرُونَ inflected, like all the tens, as a pluralis sanus.

21 أَحَدٌ وَعِشْرُونَ   إِحْدَى وَعِشْرُونَ

30 ثَلَاثُونَ, 40 أَرْبَعُونَ, 50 خَمْسُونَ, 60 سِتُّونَ

70 سَبْعُونَ, 80 ثَمَانُونَ, 90 تِسْعُونَ

100 مِائَةٌ (also written مِئَةٌ, and always so pronounced, *mi'atun*, the ا having no effect on the pronunciation).

92. THE CARDINAL NUMBERS.

500 أَرْبَعُ مِائَةٍ, 400 *ثَلَاثُ مِائَةٍ, 300 مِائَتَانِ, 200
ثَمَانِى 800, سَبْعُ مِائَةٍ 700, سِتُّ مِائَةٍ 600, خَمْسُ مِائَةٍ
مِائَةٍ, 900 تِسْعُ مِائَةٍ.

آلَافٌ) ثَلَاثَةُ آلَافٍ 3000, أَلْفَانِ 2000, أَلْفٌ 1000
is here a broken plural of the form أَفْعَالٌ § 88 No. 17)
1 000 000, مِائَةُ أَلْفٍ 100 000, أَحَدَ عَشَرَ أَلْفًا 11 000 &c.
أَلْفُ أَلْفٍ.

92. The following are the leading points to be noted in joining the cardinals to the names of the objects numbered:

a. The numerals for one (وَاحِدٌ) and two are adjectives; the numbers from 3—10, on the other hand, are substantives, and take the word indicating the objects numbered in the genitive plural. They may also, however, be placed in apposition *after* the noun. Whatever their position relative to the substantive may be—even, in fact, when the latter is altogether omitted, or when they stand as the predicate of a sentence—the construction is such that nouns of the masc. gender take the fem. forms of these numerals,

---

* Often written ثَلَاثُمِائَةٍ &c.

and *vice versâ* nouns of the fem. gender take the masc. forms. Thus: (بَنُونَ ثَلَاثَةُ) بَنِينَ three sons, (بَنَاتٍ أَرْبَعُ) أَرْبَعُ بَنَاتٍ four daughters. Also before broken plurals of which the singular is masculine, we find the fem. forms of these numerals (3—10), as ثَلَاثَةُ رِجَالٍ 3 men.

b. The numbers from 11 to 99 are followed by the word indicating the objects numbered in the *accusative singular*, as ثَلَاثُونَ رَجُلًا 30 men.

c. The numbers from 100 upwards take the thing numbered in the *genitive singular* as أَرْبَعُ مِائَةِ رَجُلٍ 400 men.

d. In the compound numbers the nature of the construction depends on the last numeral. The particle وَ is used to join the numbers together; the units and the tens may stand either before the hundreds, or after the thousands and hundreds. Thus the year 1895 is either خَمْسٌ وَتِسْعُونَ وَثَمَانِى مِائَةٍ وَأَلْفٌ or أَلْفٌ وَثَمَانِى مِائَةٍ وَخَمْسٌ وَتِسْعُونَ سَنَةً.

93. The ordinals have, for the most part, the form
a. of the act. part. of the I stem, as may be seen from the following:

93. THE ORDINAL NUMBERS.

Masc.	Fem.		Masc.	Fem.
1. أَوَّلُ, first	أُولَى	6.	سَادِسٌ	سَادِسَةٌ
2. ثَانٍ	ثَانِيَةٌ	7.	سَابِعٌ	سَابِعَةٌ
3. ثَالِثٌ	ثَالِثَةٌ	8.	ثَامِنٌ	ثَامِنَةٌ
4. رَابِعٌ	رَابِعَةٌ	9.	تَاسِعٌ	تَاسِعَةٌ
5. خَامِسٌ	خَامِسَةٌ	10.	عَاشِرٌ	عَاشِرَةٌ

11. حَادِىَ عَشَرَ   حَادِيَةَ عَشْرَةَ   indeclinable

12. ثَانِى عَشَرَ   ثَانِيَةَ عَشْرَةَ   „

13. ثَالِثَ عَشَرَ   ثَالِثَةَ عَشْرَةَ   and so on.

The ordinals of the numbers from 20 upwards are expressed by the corresponding cardinals, as ثَالِثٌ وَثَلَاثُونَ thirty-third; when larger totals have to be expressed, the cardinals are used even for the lower numbers. In dates, as a rule, the cardinal numbers are used exclusively, as فِى سَنَةِ ثَلَاثَ عَشْرَةَ وَثَلَاثِ مِائَةٍ وَأَلْفٍ مِنَ ٱلْهِجْرَةِ in the 1313th year of the Hegira (which began on the 24th of June 1895).

Fractions are usually expressed by the form فُعْلٌ, b. as ثُلْثٌ a third.

Chapter V. The Particles. (§§ 94—96).

94. The adverbs, prepositions and conjunctions cannot here be given in detail. The prepositions, like many adverbs, are still for the most part recognizable as nouns of three radicals originally, which have preserved the accusative ending without the nunation. Prepositions therefore always govern the genitive case in Arabic and may also stand in the genitive in dependence on other prepositions. Thus فَوْقَ above, with a subst. فَوْقَ ٱلْجَبَلِ up on the hill.

NOTE. A few adverbs end in *u* (which in this case has absolutely nothing to do with the nominative termination) as بَعْدُ afterwards; so مِنْ بَعْدُ in the same sense; but as prepositions بَعْدَ or مِنْ بَعْدِ after.

95. The following particles (in alphabetical order) because written with a single letter are inseparably joined to the following word, cf. § 8 note.

a. أَ (ה) interrogative particle, as أَقَتَلَ did he kill? Before the connective Alif: أَسْمُكَ for أَ + ٱسْمُكَ is thy name ... ?

b. بِ (בְּ) preposition 'in'; with suffixes thus: 1. بِى in me, 2. masc. بِكَ, 3. masc. بِهِ (§ 12*d*) &c.

c. تَ particle of asseveration, as تَٱللَّهِ by God.

سَ shortened from سَوْفَ, a particle which gives *d.*
to the impf. the sense of the future, as سَيَقْتُلُ he
will kill.

فَ, then, denotes a less close connection than وَ. *e.*

كَ (כְּ) like, as. *f.*

لَ a corroborative particle before verbs, especially *g.*
in oaths, as لَيَقْتُلَنْ he will certainly kill; it also
stands before nouns, especially after the particle إِنَّ
(§ 125 *a* note).

لِ (לְ) preposition and conjunction; before suffixes *h.*
(except in 1. pers. sing. لِي) it becomes لَ, as لَكَ
to thee.

وَ (ו, וּ) connective particle; as a particle of *i.*
asseveration it takes the gen., as وَٱللّٰهِ by God.

As regards the addition of pronominal suffixes 96.
to the prepositions and conjunctions, the following
points may be noted in addition to what has been
said under § 82.

Before the suffixes of the 1. pers. sing., the final *a.*
vowel or vocalic *auslaut* is dropped as is the case
with the noun; thus بَعْدَ 'after' with the suff. of the
1. pers. sing. بَعْدِى, but بَعْدَكَ &c.

b. In the prepositions عَلَى upon, and إِلَى towards, the final ى is sounded before suffixes (contrary to § 2 *d*), e. g.

with suff. of the 2. pers. masc. إِلَيْكَ, عَلَيْكَ

„ „ „ 3. „ „ إِلَيْهِ, عَلَيْهِ

„ „ „ 1. „ „ إِلَىَّ, عَلَىَّ (see § 82 *a*)

c. The prepositions مِنْ and عَنْ double the *n* before the suffix of the 1. pers. sing., as مِنِّى.

d. إِنَّ behold, truly, and أَنَّ that, become with the suff. of the 2. pers. sing. masc. إِنَّكَ and أَنَّكَ

„ „ „ „ 1. „ „ إِنِّى, or إِنَّنِى

أَنِّى, or أَنَّنِى

„ „ „ „ 1. „ plur. إِنَّا or إِنَّنَا

أَنَّا or أَنَّنَا.

III. NOTES ON SYNTAX. (§§ 97—160).

Chap. I. Moods and Tenses. (§§ 97—104).

97. The *perfect* expresses a completed action, the completion of which falls in the past, present or future, or is thought of as falling in one or other of these

periods. The *imperfect* expresses an uncompleted action, which may likewise fall in each of the same three spheres of time.

The perfect is, in the first place, the tense of narration (perfectum historicum), when an action completed in the past is spoken of, and may, as a rule, be rendered by our past tense, as جَاءَ زَيْدٌ Zaid came. [98. a.]

By the perfect the idea is expressed that an action or a state has continued from the beginning, and still continues, as إِخْتَلَفُوا ٱلْعُلَمَاءُ the learned (always) disagree (gnomic aorist); ٱللّٰهُ تَعَالَى God, he is exalted (from the beginning). [b.]

When the perfect expresses an action completed in the present, it is to be rendered by our present, as أَعْطَيْتُكَ هٰذَا I present you with this (the affair is at this moment concluded). [c.]

In a sentence containing an oath or a wish, the perfect expresses an action which, in the mind of the speaker, is completed in the future, as لَعَنَهُ ٱللّٰهُ God curse him; also with لَا 'not', as لَا رَحِمَهُ ٱللّٰهُ may God have no pity on him; وَٱللّٰهِ لَا فَعَلْتُ by God I do it not! [d.]

When the particle قَدْ stands before the perfect, the latter may in most cases be rendered by our per- [e.]

fect (either the present or the past perfect), as قَدْ ذَكَرْنَا we have (just) mentioned, or we had mentioned. The perf. with قَدْ may also be used in the sense given under sub-section c.

f. When the verb كَانَ (to be) stands before the perfect (with or without قَدْ), we must render as a rule by our past perfect (pluperfect), as لَمَّا وُلِدَ مُوسَى كَانَ قَدْ أَمَرَ فِرْعَوْنُ بِقَتْلِ ٱلْأَطْفَالِ when Moses was born, Pharaoh had (just) commanded to kill the little children.

NOTE. Instead of the above verbal sentence (§ 134), كَانَ may be followed by a compound nominal sentence (§ 138 d) as كَانَ فِرْعَوْنُ قَدْ أَمَرَ ...

g. Our conditional is expressed in Arabic by the perfect, that is, it is represented as something already accomplished, as وَدَدْتُ I should wish, (قَدْ) كُنْتُ وَدَدْتُ I should have wished.

h. For the perf. after إِذَا and in conditional sentences see §§ 157, 158.

99. The *imperfect* indicative is to be rendered according to circumstances by our present or our future, sometimes also by our past progressive (imperfect).

a. If the future is to be expressed with greater precision than by the Arabic imperfect alone, the latter

99. THE IMPERFECT.

has prefixed to it the adverb سَوْفَ (end), which may be shortened to سَ and is then inseparably joined to the verb (see § 95 d), as سَوْفَ تَعْلَمُونَ ye will know (it); سَنُرِيهِمْ (49 b) we shall show you.

By the imperfect is expressed an action which *b.* accompanies another action completed in the past, or which is still in the future from the stand point of the latter, as جَآءُوا أَبَاهُمْ يَبْكُونَ they came to their father weeping (cf. § 157 b); أَتَى ٱلْعَيْنَ يَشْرَبُ he came to the spring to drink.

The imperfect can also express the continuance *c.* of an action in the past; يَتَقَاتَلُونَ may also mean 'they were fighting for a considerable time', or 'they fought repeatedly, with each other'. More frequently, however, this continuous imperfect is expressed by a combination of كَانَ with the impf. (cf. § 98 f. and note); sometimes we can render such a combination by our 'was wont to' or 'used to', as كَانَ يَأْخُذُ فِي كُلِّ يَوْمٍ ثَلَاثَةَ دَرَاهِمَ he used to receive every day three drachmae.

If قَدْ stands before the imperfect, a certain in- *d.* definiteness is the result, as قَدْ يَكُونُ 'it will most likely be that ...', an idea which is not unfrequently found in the imperf. without قَدْ.

NOTE a. The impf. also stands in direct subordination to other verbs, as مَا زِلْتُ أَشْرَبُ I ceased not to drink (cf. § 110); جَعَلَ يُكَلِّمُ ٱلنَّاسَ he began to speak with the people; مَا أَقْدِرُ أَفْعَلُ كَذَا I cannot do such a thing.

NOTE b. Before several verbs (perfects or imperfects) joined together with وَ, it is sufficient to write كَانَ once, and so with قَدْ, سَوْفَ and سَ.

NOTE c. كَانَ (see note to § 98*f*) is frequently followed by a compound nominal sentence, as كَانَ عُثْمَانُ يَزُورُ ٱلْمَقَابِرَ Osmān was wont to visit the graves (the cemetery).

100. The *Subjunctive* is found in certain kinds of dependent clauses introduced by a conjunction, the action of which is to be represented as one to be expected as the result of the action of the principal clause, and hence as one that is only likely to occur in the future. Hence this mood is frequently (not always) used after the conjunctions أَنْ that, أَلَّا (from أَنْ لَا) that not, حَتَّىٰ until, فَ (and وَ) that, and always after (لِأَنْ لَا) لِ, كَيْ, لِأَنْ in order that, لِئَلَّا (made up of لِأَنْ لَا) in order that .. not, أَوْ in the sense of 'except that', 'until', as جَآءَ لِيَزُورَنِي he came in order to visit me; أَمَرَهُ أَنْ يَكْتُبَ he commanded him to write (that he

should write). In like manner the subj. is used after لَنْ (أَنْ لَا) it will not be (the case) that, as لَنْ أُرْسِلَهُ I shall not send him.

The *modus apocopatus* (or jussive) is found: **101.**

1) in positive commands, generally with the particle *a.* لِ prefixed, as لِيَكْتُبْ let him write.

NOTE. When such a form is further preceded by وَ and فَ (which is sometimes the case, without any special stress resting on these particles) لِ generally loses its vowel, as وَعَلَى ٱللّٰهِ فَلْيَتَوَكَّلِ ٱلْمُؤْمِنُونَ and in God let the believers (then, therefore) trust.

2) in negative commands with لَا, as لَا تَقُلْ say not, *b.* thou shalt not say. The imperative can never take a negative.

3) always after لَمْ, not as a prohibition but as ne- *c.* gativing a completed action, as لَمْ يَضْرِبْ he did not strike, (as the negation of ضَرَبَ); in like manner after لَمَّا in the sense of 'not yet'.

4) in the protasis and apodosis of conditional sen- *d.* tences, see § 158 ff.

The *modus energicus* is usually found in assevera- **102.** tions, and particularly in connection with an oath and the corroborative particle لَ, as وَٱللّٰهِ لَأَضْرِبَنَّهُ by God, I will certainly strike him; this mood is also used with the prohibitive لَا.

103. The *Passive* is employed in those cases in which the agent, for some reason or other, must not be mentioned. Hence a sentence like قُتِلَ زَيْدٌ means 'Zaid has been killed (by some person unknown or who may not be named)'. Our 'Zaid has been killed by ʻAmr', the Arabs express by the active construction. The passive is frequently found in an impersonal sense (see § 121 a).

104. With regard to the employment of the *participles* the following points are to be noted:

a. The participle (especially as predicate of a nominal sentence § 122 a) frequently expresses our "to be about to", as أَنَا قَادِمٌ إِلَيْكَ I am about to come, on the point of coming, to you.

b. The passive participle is also used impersonally in Arabic; starting from the sentence غُشِيَ عَلَيْهِ he fainted (literally: it was covered over him) we can also say هُوَ مَغْشِيٌّ عَلَيْهِ he has fainted, fem. هِيَ مَغْشِيٌّ عَلَيْهَا. In such constructions the impersonal part. pass. may be inflected for all three cases and be determined by the article, as مَرَرْتُ بِرَجُلٍ مَغْشِيٍّ عَلَيْهِ I passed a man who had fainted; رَأَيْتُ ٱلْمَرْأَةَ ٱلْمَغْشِيَّ عَلَيْهَا I saw the woman that had fainted.

Chap. II. The Government of the Verb. (§§ 105—117).

105. In Arabic the verb may take as its complement either an accusative, or a preposition with its case. The numerous combinations of the latter sort, in which the preposition with its case is sometimes the necessary complement of the action denoted by the verb, sometimes merely accessory (such, for example, as specifications of place and time) cannot here be given in detail. See, however, §§ 114 ff.

106. The *accusative* is the case depending immediately on the verb. We distinguish here the cases in which, the accusative stands α) as object, β) as predicate, and γ) as limitation or more precise definition, generally called by grammarians, the accusative "of nearer definition".

107. α) Certain classes of verbs, as for example, verbs of coming and going, take as direct object the goal to which the action is directed, e. g. دَخَلَ ٱلْبَيْتَ he went into the house.

NOTE. On the other hand دَخَلَ إِلَى ٱلْبَيْتِ denotes primarily the *direction* of the action *towards* the goal; دَخَلَ فِى ٱلْبَيْتِ he went into the house and stayed there.

108. The following take *two* accusatives: 1) The causative forms of transitive verbs with one accusative in the I. stem, as عَلِمَ to know; caus. عَلَّمَهُ ٱلْقِرَآءَةَ he taught him reading; 2) verbs that express the ideas of filling or giving, of making into, of considering or recognising as, of naming, and many others: e. g. جَعَلَ ٱللّٰهُ ٱلْأَرْضَ فِرَاشًا God made the earth (into) a carpet; سَمَّى ٱبْنَهُ مُحَمَّدًا he named his son Muhammed. When a verb of this class is put in the passive, the second accusative remains, as سُمِّى ٱبْنُهُ مُحَمَّدًا his son was named Muhammed; أُوتِىَ دِرْهَمًا he was presented with a dirhem, from the active آتَاهُ دِرْهَمًا he presented him with a dirhem (for suff. see § 107).

NOTE a. The two accusatives of such verbs as express the idea of finding one to be, or considering one as something, stand to each other, strictly speaking, in the relation of subject and predicate (§ 139); thus a sentence like وَجَدْتُهُ شَيْخًا حَلِيمًا may also be translated 'I found that he was a gentle old man'. As second object we may have a verb instead of a noun, as وَجَدُوا بِضَاعَتَهُمْ رُدَّتْ إِلَيْهِمْ they found their payment to be something which was returned to them = they found that their payment was &c.

NOTE b. Verbs expressing not an intellectual but a physical perception are also frequently found with two accusatives. The second, indeed, is generally regarded as an acc. of condition (§ 113 b), but sentences like سَمِعْتُ عَمْرًا بَاكِيًا, it must be admitted, may also be translated: I heard ʿAmr weeping, i. e. I heard how ʿAmr wept.

109. For the purpose of strengthening or of more precisely defining the idea conveyed by it, every verb

may take a so-called *absolute object*. This absolute (or *internal*) object consists of an infinitive, a nomen speciei (§ 64 *c*) or other noun. Usually this object is itself more precisely defined either by some qualifying word or phrase (§ 120) or by a genitive, as أَدَّبَهُ تَأْدِيبًا حَسَنًا he educated him with a good education, i. e. well; ضَرَبَنِى ضَرْبًا أَوْجَعَنِى he struck him with a stroke which pained me (for the relative sentence, see § 155); سَلَكَ سِيرَةَ جَدِّهِ he walked in the way of his grand father. More rarely the absolute object is found without any qualification, as ضَرَبَهُ ضَرْبًا he struck him with a stroke, as much as to say, he struck him a blow, and what a blow! صَرَّهُ صُرَرًا he wrapped it in (so many) parcels; here the absolute object expresses rather the result of the action.

NOTE. Sometimes the place of the infinitive is taken by the mere qualification, as سَارَ سَيْرًا طَوِيلًا = سَارَ طَوِيلًا he journeyed long, for he journeyed a long journey, or by some other form of nearer definition, as فَتَحَ ٱللّٰهُ عَلَيْهِ بَيْتَ ٱلْمَقْدِسِ صُلْحًا God allowed him to capture Jerusalem peacefully = فَتَحَ صُلْحًا.

β) The accusative stands as the *predicate* with verbs which express the idea of being or becoming something, and is especially common with the verb كَانَ (med. و). This verb signifies either 1) to be in the

7*

sense of to exist, as كَانَ وَزِيرٌ there was (there lived) a vizier, or 2) to be something (in particular); in the latter sense it takes its predicate (to adopt the nomenclature of the native grammarians) in the accusative, as كَانَتِ ٱمْرَأَتُهُ حَامِلًا his wife was pregnant. The same construction is adopted by all verbs of similar signification, such as أَمْسَى to be something late, أَصْبَحَ to be something early, عَادَ to be or become something a second time, دَامَ to remain, to last, زَالَ to cease to be something, صَارَ to become something, لَيْسَ not to be something. The place of the accusative in the predicate may be taken by a preposition with its case (cf. § 114 ff.), as كَانَ زَيْدٌ فِي ٱلْبَيْتِ كَانَتْ مُلُوكُ ٱلْفُرْسِ Zaid was in the house; مِنْ أَعْظَمِ مُلُوكِ ٱلْأَرْضِ the kings of Persia belonged to the most powerful sovereigns on earth. The construction of كَانَ and the others with a finite verb (§§ 98*f*; 99*c*) must also be understood in this way, that is, the predicate in such cases consists of a verbal sentence (§ 135), as أَصْبَحَ ٱلنَّاسُ قَدْ تَعِبُوا the people had already (prop. early) become weary.

The accusative, further, stands in the predicate

after the negative لَا, when the latter, as the Arabs say, expresses a *general* negation. The accus. after لَا, which is always undetermined, drops its nunation, as لَا إِلٰهَ إِلَّا ٱللّٰهُ there is (absolutely) no God but Allah.

112. The accusative is used after the conjunction وَ to indicate concomitance, especially in verbal sentences (§ 135), as مَا صَنَعْتَ وَأَبَاكَ what hast thou and thy father done? مَا زِلْتُ أَسِيرُ وَٱلنِّيلَ I ceased not to go with (along) the Nile; also without a verb مَا لَكَ وَزَيْدًا what hast thou (to do) with Zaid?

113. ٧) The *accusative* of *nearer definition* is employed in the following cases:

a. 1) To give details of place and time, as نَظَرَ يَمِينًا وَشِمَالًا he looked to right and to left of him; سَارَ فَرْسَخًا he journeyed a parasang; جَاؤُا عِشَاءً they came late in the evening; اِسْتَمَرَّ عَلَى ذٰلِكَ مُدَّةَ حَيَاتِهِ he continued faithful thereto during his life-time.

b. 2) Very frequently the accusative, as a rule undetermined, appears in verbal (rarely in nominal) sentences as the accusative of state or condition, as سَارَ مُتَوَجِّهًا إِلَى ٱلْمَدِينَةِ he journeyed, taking the direction of Medina; لَقِيتُ عَمْرًا بَاكِيًا I met 'Amr weeping.

NOTE a. With the accusative of condition the student must be careful to note to which of the nouns in the sentence it applies; in the last sentence above, for example, it might refer to the subject pronoun implicit in لَقِيتُ instead of to ʿAmr.

NOTE b. Two nouns in the accusative of condition are often placed beside each other without a conjunction (asyndeton) as اخْرُجْ مِنْهَا مَذْوُمًا مَدْحُورًا (God said to Satan): Go out of it (paradise, fem.) as one cast off and despised (for مَذْوُمًا see § 7 b note).

NOTE c. In some rare cases an infinitive is used (in place of a participle) to denote a qualifying circumstance; قُتِلَ صَبْرًا he was killed bound (i. e. while bound) = مَقْبُورًا.

c. 3) The accusative of *specification* (= accus. of respect), also in most cases undetermined, expresses a more precise reference, as حَسُنَتْ مُسْتَقَرًّا it (paradise) is beautiful with reference to staying (there), i. e. as a dwellingplace; this accus. is especially common with elatives (§ 63 b) of a more general signification, as أَشَدُّ حُمْرَةً stronger with regard to the colour red = redder.

d. 4) The accusative of nearer definition is also employed to indicate the *motive* or *purpose* of an action, in which case, also, it is mostly undetermined, as هَرَبُوا إِكْرَامًا لَهُ they fled from cowardice; قُمْتُ إِكْرَامًا لَهُ I stood up to do him honour.

114* The accusative may also stand in cases, particularly in exclamations, where a finite verb can be supplied, as أَهْلًا وَسَهْلًا welcome! Here we must supply جِئْتَ, and the meaning of the phrase comes to be: thou art come to relatives and a smooth (i. e. pleasant) place; مَهْلًا slowly! to be taken as the absolute object of an imperative understood.

Of the numerous constructions of the *verb with a preposition* attention need only be called to the following. **114.**

Many prepositions are still treated as nouns, in accordance with their original signification (see § 94), as مَيَّزَ بَيْنَ ٱلذُّكُورِ وَٱلْأَنَاثِى he distinguished between (prop. the distance, difference of) males and females. Very frequently we find (cf. § 110) the partitive مِنْ used in this way as object, e. g. أَكَلَ مِنَ ٱلطَّعَامِ he ate of the food.

A few verbs are construed, with but slight differ- **115.** ence of meaning, now with a direct object, now with بِ, as عَلِمَهُ he knew it, عَلِمَ بِهِ he knew about it. Frequently بِ serves to introduce an object, to which the action of the verb extends only indirectly, as بَعَثَ زَيْدًا he sent Zaid; بَعَثَ بِٱلْكِتَابِ he sent the writing (i. e. some one with the writing); بَعَثَ ٱلْعَبْدَ he sent the slave, بَعَثَ بِٱلْعَبْدِ, same meaning, but with the understanding that the slave travels under escort. Verbs of going construed with بِ take the sense of bringing, as أَتَى زَيْدًا بِٱلْخَبَرِ he brought Zaid the news.—This بِ may also accompany an imperative as a periphrasis of the first person of the dual and plural, as اِمْضِ بِنَا let (thou) us go, اِمْضُوا بِنَا let (ye) us go.

116. The meaning of many verbs is often so altered according to the preposition with which they are construed that a sense quite the opposite of the original, according to our idiom, is the result; thus دَعَا لَهُ is properly: he called (to God) in his favour, i. e. he blessed him, دَعَا عَلَيْهِ he called (to God) against him, i. e. he cursed him; اِشْتَغَلَ بِٱلْأَمْرِ he occupied himself with the affair; but with عَنْ (which contains the idea of separation) اِشْتَغَلَ عَنِ ٱلْأَمْرِ he was occupied so that he put the affair in question aside, could not attend to it.

117. Of the various uses of the preposition لِ (see §§ 130 ff.), we may call attention to its special use in dates, particularly in specifying the days of the month, as لِأَوَّلِ لَيْلَةٍ مِنْ مُحَرَّمٍ in the first (literally: to the first) night of (the month) Muḥarram. لِسَبْعِ لَيَالٍ خَلَوْنَ

مِنْ شَعْبَانَ or with the omission of لَيَالٍ (§ 90 p) لِسَبْعَ خَلَوْنَ at the time of seven nights, which (cf. § 155) had elapsed of Šaʿbān, i. e. when seven nights (or days) of Š. had passed; بَقِينَ (لَيَالٍ) لِأَرْبَعَ عَشْرَةَ مِنْ رَمَضَانَ when still fourteen (nights) were left of Ramaḍān.

Chap. III. The Government of the Noun. (§§ 118—134).

118. A noun may take with it α) the article, β) a permutative (noun in apposition), γ) a qualifying (attributive) adjunct, δ) a genitive.

α) When a noun is preceded by the *article*, it is said to be *determined* (§ 79 b). This determination may be stronger or weaker:

a. A very strong determination is found in certain words which contain the idea of time, as اَلسَّاعَةَ this hour = now, اَلْيَوْمَ this day = today. In these cases the article has the force of a demonstrative.

b. By means of the article a single definite object is indicated, which the speaker has in mind, or which has been already mentioned: by اَلرَّجُلُ is meant some particular known man. Proper names furnished with the article (see § 79 a) were originally appellatives with the determination, as اَلْحَسَنُ.

c. The determination by the article often serves merely to denote the *species* or class to which something belongs, as هُوَ مِثْلُ ٱلْحِمَارِ he is like an ass. This use of the article is named the generic.

119. β) From among the cases in which a noun follows another noun in *apposition*, the following may be singled out as worthy of note:

a. A substantive may have in apposition words expressing a) size, b) resemblance, c) the parts and d) the material of which a thing is made up. Thus a) ثَوْبٌ ذِرَاعٌ a dress an ell long (lit. a dress, an ell); b) رَجُلٌ مِثْلُ زَيْدٍ a man like (lit. the likeness of) Zaid; c) حَبْلٌ أَرْمَاتٌ a rope made up of rotten pieces; d) ٱلْخَاتَمُ ٱلْحَدِيدُ the iron finger-ring; when undetermined preferably with مِنْ as صَنَمٌ مِنْ ذَهَبٍ an idol of gold. For the last, the genitive construction is also found viz: صَنَمُ ٱلذَّهَبِ.

b. The word كُلّ totality is construed either with the noun following in the genitive, or stands in apposition, with a suffix referring back to the noun, as كُلُّ ٱلنَّاسِ or ٱلنَّاسُ كُلُّهُمْ all men. (Note that كُلّ being a substantive always remains unchanged as regards gender and number).

γ) A substantive may be *qualified* 1) by an adjective, 2) by a preposition with its case, or 3) by a relative clause (§§ 155—6).

1) The qualifying word may be an *adjective*, as إِمَامٌ عَادِلٌ an honest Imām; in this case if the substantive is determined the adjective must also receive the determination, as اَلْإِمَامُ ٱلْعَادِلُ, the honest Imām.

The adjective *follows* its substantive; to this rule the demonstrative pronoun forms an apparent exception, in as much as it generally stands *before* (like the article § 118), less frequently *after*, the substantive which it qualifies. Thus we find هٰذَا ٱلْغُلَامُ this slave, alongside of اَلْغُلَامُ هٰذَا.

The adjective must agree with its substantive in gender and number, as صَبِيَّةٌ جَمِيلَةٌ a pretty girl. Among the exceptions is the word كَثِيرٌ much, which generally remains unchanged, like a noun in apposition, even after the plural, as رِجَالٌ كَثِيرٌ, many men.

That the broken plurals take their adjectives in the feminine has been already noted (see § 87 a); the adjective, however, may also take a broken plural, as رِجَالٌ كِرَامٌ, noble men. The plur. sanus, moreover, is not unfrequently found especially if the adjective qualifies words denoting living beings, as اَلْآبَآءُ ٱلْمَاضُونَ the ancestors that were of old (part. of مَضَى). In the same circumstances the collectives (§ 86 *a*) may also take a plural adjective, as قَوْمٌ بُخَلَاءٌ miserly people, قَوْمٌ ظَالِمُونَ violent people. The preceding pronoun often stands then in the plural, as هٰؤُلَآءِ ٱلنَّاسُ these men; but with fem. plurals that do not denote living beings generally in the fem. singular, as هٰذِهِ ٱلْفَلَوَاتُ these deserts; before broken plurals also in the fem. sing., as هٰذِهِ ٱلْمَمَالِيكُ these slaves.

121.
a. 2) From those cases in which a *preposition with its noun* is dependent on a verb (§§ 114 ff.) or its equivalent, must be clearly distinguished those in which they form the qualifying attribute of another noun, as جَلَسْتُ عَلَى صَائِغٍ بِٱلسُّوقِ I sat down beside a goldsmith (who was) in the bazaar; ذُرِّيَّتُكَ مِنْ بَعْدِكَ thy posterity (that will be) after thee.

b. Sometimes this attribute does not stand next to the word qualified; so particularly with the relatives مَنْ and مَا, as مَنْ دَخَلَ ٱلشَّأْمَ مِنَ ٱلْعَرَبِ those of the Arabs that advanced into Syria; اَنْكِحُوا مَا طَابَ

لَكُمْ مِنَ ٱلنِّسَاءِ marry of the women whatever seemeth good unto you.

122. Should several attributes qualify a single substantive, the connecting conjunction is usually omitted (asyndeton), as اَللّٰهُ ٱلْعَلِىُّ ٱلْعَظِيمُ the high and mighty God; بَابٌ وَاسِعٌ مِنْ أَبْوَابِ ٱلْقَرْيَةِ a wide gate of the gates of the town; غَمَامَةٌ ضَخْمَةٌ عَلَى رَأْسِهِ تُظِلُّهُ a thick cloud over his head which gave him shade.

123. δ) One noun, when in dependence on another, is put in the *genitive* case — the function of which is to determine more exactly the application of the preceding noun. As the result of the close connection subsisting between the second noun and the first, the latter, now said to be in the construct state (§ 79 c) and therefore without the article, is regarded as *determined*. Therefore رُمْحُ ٱلْفَارِسِ is 'the (particular) spear of the (particular) horseman', and so with the suffixes, as رُمْحُهُ his (particular) spear. When the dependent noun (nomen rectum) is undetermined, the governing noun (nomen regens) is only defined in a generic sense (§ 118 c), or is specialized in a way resembling the generic definition, as بِنْتُ مَلِكٍ a daughter of a king = a king's daughter.

Note. More rarely, in the latter case, the generic article may be attached to the nomen rectum, as خُبْزُ ٱلشَّعِيرِ barley bread

124. The genitive cannot be separated from the governing word (nomen regens); adjectival and other additions must therefore stand *after* the genitive, as بَيْتُ ٱلْمَلِكِ ٱلْوَاسِعُ the spacious house of the king. When, according to our idiom, a genitive belongs to two substantives, in Arabic it is made dependent on the first of the two, and represented with the second by a personal pronoun, as رَحْمَةُ ٱللّٰهِ وَبَرَكَاتُهُ the mercy and blessings of God.

125. Substantives conveying the idea of time sometimes receive a specially strong determination by the addition of suffixes (cf. § 118 a), as صَلَّى لَيْلَهُ he prayed his night, i. e. the particular night in which he then was.

126. The close connection of two nouns thus standing in the genit. relation makes sometimes possible their fusion to *one* idea, although only the first component admits of inflection. Thus عَبْدُ ٱللّٰهِ (gen. عَبْدِ ٱللّٰهِ; acc. عَبْدَ ٱللّٰهِ) the servant of Allah, as a proper name, conveys but a single idea. Further illustrations will be found in the numerous examples of composite proper names, of which one of the elements is one or other of the words ٱبْن son, أَب father, بِنْت daughter, أُم mother.

As the Arabs have no family names, properly so-called, the name of a man or woman receives for distinction's sake an addition by the help of the above words, as أَبُو ٱلْعَبَّاسِ مُحَمَّدُ بْنُ يَزِيدَ (observe the order). Very frequently a name thus made up has become the principal name, as that of the first Caliph أَبُو بَكْرٍ, for example, or that of the savant اِبْنُ قُتَيْبَةَ; names of tribes, too, like بَنُو كَلْبٍ, are in the same way simple notions (Einheitsbegriffe).

127. Not unfrequently an adjective which in our idiom would be made to qualify its substantive, is in Arabic raised to the rank of a substantive, on which its proper substantive is made to depend; thus كَرِيمُ خُلُقِهِ the noble(ness) of his character = his noble character; أَكْثَرُ ٱلنَّاسِ most men. The same construction is found with elatives also, as عَاشُوا أَهْوَنَ عِيشَةٍ they lived the easiest life (cf. § 109).

128. A species of explicative genitive is found in cases where a general conception is more explicitly defined by a following proper name, as أَرْضُ ٱلْيَمَنِ the land of Yemen. — Under this head may be reckoned the suffixes appended to numerals, as ثَلٰثَتُهُمْ the three of them.

129. A few words containing the ideas of time and place may have, instead of a genitive, a whole clause depending on them, as يَوْمَ قُتِلَ on the day on which he was killed.

130. When a noun on which another noun is in the proper sense (cf. § 134) dependent must remain absolutely undetermined (see § 123), the usual genitive relation of nomen regens and nomen rectum is inadmissible, and the connection of the two must be expressed by a preposition, as أَخٌ لَكُمْ a brother of yours, where لَكُمْ is attrib. adjunct to أَخٌ (see § 121 a).

131. *Infinitives* may govern their object according to the laws either of verbal or of nominal government. In the first instance their *subject* is subordinated in the genitive; قَتْلُ زَيْدٍ accordingly means: the circumstance that Zaid has killed. If no subject is named, the object may likewise stand in the genitive, so that the same expression قَتْلُ زَيْدٍ may also mean: the circumstance that Zaid has been killed, the fact of Zaid's being killed. When both subject and object are present, the former is treated as a subjective genitive; the latter remains in the accusative or ل with the genitive is used as a periphrasis for the accusative, as شُرْبُ الْخَمْرِ مُدَامَوَمَتْهُ

the circumstance that he was constantly drinking wine; حَتَّى لِلْخَمْرِ the circumstance that I am fond of wine. لِ also stands after an undetermined infinitive (e. g. in cases like § 113 d and others) as قُمْتُ إِكْرَامًا لِزَيْدٍ I stood up to do honour to Zaid.

132. In the case of the participle, the object of the verb appears as the objective genitive, and when the part. has the sense of the perfect it is determined by the genitive following, as اَللهُ خَالِقُ ٱلْأَرْضِ God is he who has created the earth = the creator of the earth. With a present or future sense the governing participle is not determined, as كُلُّ نَفْسٍ ذَائِقَةُ ٱلْمَوْتِ every soul is one that will taste of death; إِنَّهُ مُلَاقِيكُمْ he is one that will meet with you. If the participle is in itself determined, the object stands in the accusative or is expressed periphrastically with لِ, as اَلطَّالِبُ لِلْعِلْمِ he who strives after knowledge; the same applies when the participle is strictly undetermined, as طَالِبٌ ثَأْرَ أَبِيهِ one who wishes to take blood revenge for his father; مَا زِلْتُ مُحِبًّا لِلْإِسْلَامِ I have not ceased to love Islam.

133. A special idiomatic use of certain generic words is their combination with a following genitive. They are determined or undetermined according to the context, e. g. ذُو he who has, possessor of (cf. § 90*l*), ذُو مَالٍ the possessor of wealth, a rich man; صَاحِبٌ companion, owner, صَاحِبُ عَقْلٍ the man of sense; أَهْلُ people, أَهْلُ ٱلدُّنْيَا people of the world = worldly people; بَعْضٌ portion, e. g. بَعْضُ ٱلْعُلَمَآءِ one, some of the learned; غَيْرٌ prop. change, then 'another than', as مَاتَ ٱلْمَلِكُ فَمَلَكَ بَعْدَهُ غَيْرُهُ the king died and another than he reigned after him; similarly أَحَدٌ one, as أَحَدُهُمْ one of them; finally ٱبْنٌ son, in certain common idioms, as ٱبْنُ ثَلَاثِينَ سَنَةً thirty years old.

134. A special kind of genitive relation is presented by the so-called *improper* annexation, by which a participle or a verbal adjective (see § 60*b*) is more strictly limited or defined by a following genitive, as رَجُلٌ حَسَنُ ٱلْوَجْهِ a man beautiful of countenance. This construction is best rendered by a relative clause, the subject of which will be the word that more clearly defines the governing idea, in other words the genitive of the Arabic will be the nominative of the English, a man whose countenance is beautiful. In such a case the governing word is *not* determined by the following genitive; should the latter require to be determined, it may receive the *article* (contrary to the rule in § 123) as اَلرَّجُلُ ٱلْحَسَنُ ٱلْوَجْهِ the man of the beautiful countenance, i. e. whose countenance is beautiful.

Chapter IV. The Simple Sentence. (§§ 135—151).

135. Sentences in Arabic are of two kinds, *verbal* and *nominal*.

The chief characteristic of a *verbal sentence* is the fact that it always contains a finite verb; in fact, a verb of this kind with its inherent (subject) pronoun is in itself a complete verbal sentence, as ضَرَبْتَ thou hast struck. This type of sentence always expresses the *commencement of some activity*, understood in the widest sense. If a special exponent of the idea conveyed by the subject of the verb is added, it *follows* the verb in the case appropriate to the subject, *viz.* the nominative, as ضَرَبَ زَيْدٌ he has struck, Zaid (has) = Zaid has struck, whereby Zaid is singled out as the agent.

136. In the *verbal sentence*, the finite verb does not always agree in gender and number with the following subject. The following are the chief points to be noted in this connection:

a. The verb stands in the *masculine singular* before sound or outer plurals, and generally before the masc. forms of the dual.

b. The verb stands in the *feminine singular* 1) before a sing. fem. if it follows the verb immediately, 2) before sound plurals feminine, 3) before the fem. forms of the dual, and 4) before broken plurals (cf. next sub-section).

c. The verb stands in the *masculine or feminine singular* 1) before a sing. fem. not immediately following the verb, 2) before collectives, 3) before broken plurals denoting male persons; if these plurals do not immediately follow the verb, the latter in most cases takes the masc. singular form.

d. Once the subject is introduced, the verbs following agree with it in gender and number, as جَاءَ زَيْدٌ وَخَالِدٌ وَعَبْدُ ٱللّٰهِ وَقَالُوا there came Zaid, Hālid and ʿAbdallah and they said. After collectives also the verb, in such a case, often takes the plural, as مَضَتِ ٱلْغِلْمَانُ يَتْبَعُونَهُ the young people set out to follow him. So too after words like قَوْمٌ and others. Still it is always possible for the verb to remain in the singular, as وَقُرَيْشٌ

137. INDEFINITE SUBJECT.

تَحْبِسُ مَنْ قَدَرَتْ عَلَى حَبْسِهِ and the Ḳuraishites (the tribe Ḳuraish) imprisoned whomsoever they could imprison.

A subject unknown, or purposely left unnamed, **137.** is treated as follows (cf. French *on dit*, German *man sagt*):

1) The verb is put in the 3. pers. sing. of the *a.* passive (see § 103), as يُسَارُ إِلَيْهِ they journey to him. It is to be noted that this impersonal passive can never stand without a complement (here إِلَيْهِ).

2) Or in the 3. pers. plur. of the active, as قَالُوا *b.* they said.

3) Or in the 2. pers. sing. (or plur.) of the active, *c.* e. g. in the Ḳur'ān أَرَأَيْتَ or أَرَأَيْتُمْ dost thou think? do ye think? where it is not any particular persons that are addressed, but people in general, as much as to say 'could any one suppose that....?' تَقُولُ one might say (cf. Eng. 'as you might say').

4) There may be added to the verb a subject *d.* (participle) formed from the same root, as قَالَ قَائِلٌ or قَالَ ٱلْقَائِلُ some one said; قَصْرٌ لَمْ يُرَوْا ٱلرَّاوُونَ مِثْلَهُ a castle, the like of which had never been seen.

NOTE. The case of an undefined complement of a verbal action being expressed by a substantive derived from the verb is not unfrequently met elsewhere than in the above construction, e. g. لَا يَخَافُونَ لَوْمَةَ لَائِمٍ قَتَلَ قَتِيلًا aliquem (interfectum) interfecit, they did not fear the reproof of any reprover.

138. Occasionally, out of something that has been mentioned, a story or the like, there arises an indefinite subject corresponding to our "it", which is usually expressed by the feminine of the verb; for example, after a fable or the like, فَذَهَبَتْ مَثَلًا, and it (i. e. this story) passed into a proverb.

139. The *nominal sentence*, in contrast to the verbal sentence, expresses a *state* or *condition* of the subject. This last as a rule stands at the head of the sentence in the case appropriate to the subject, viz. the nominative; in most cases it is determined while the predicate is undetermined. The predicate may consist of one or other of the following:

a) a simple noun, as زَيْدٌ عَالِمٌ Zaid is wise;

b) a preposition and its case, as اَلرَّجُلُ فِي ٱلدَّارِ the man is in the house;

c) an adverb, as عَبْدُ ٱللّٰهِ هٰهُنَا 'Abdallah is here.

d) a complete sentence, which may be either α) a verbal sentence, or β) a nominal sentence; the whole now becomes a compound sentence. Exx.: *a*) زَيْدٌ مَرِضَ Zaid (he) is ill; زَيْدٌ مَرِضَ أَبُوهُ Zaid, his father is ill; β) زَيْدٌ أَبُوهُ مُسِنٌّ Zaid, his father is aged (i. e. Zaid's father &c.). The sentence constituting the predicate must contain a pronoun referring back to the subject. The subj. thus placed at the head of the sentence has been wrongly named the nominative absolute.

NOTE. The difficulty we feel in distinguishing between ضَرَبَ زَيْدٌ and زَيْدٌ ضَرَبَ Zaid has struck, may be explained in this way. In the first of these two expressions it is the act of striking that is uppermost in the speaker's mind, and the enquiry as to the subject or agent from whom the act proceeds is answered with Zaid, on which the logical emphasis now rests. In زَيْدٌ ضَرَبَ, on the other hand, we start with Zaid as a given subject or agent, and the question as to what is to be predicated regarding this subject or as to what this agent has done is answered by ضَرَبَ, on which in its turn the logical centre of gravity, so to say, comes to rest.

140. Between subject and predicate, when both are determined, there ought to stand the pronoun of the 3. person, but this rule is not always observed, as اَللّٰهُ هُوَ ٱلْحَىُّ God is the living One.—Sometimes, also, this pron. merely serves to emphasize the subject.

141. In negative and interrogative sentences the predicate stands before the subject, as أَيْنَ زَيْدٌ where is Zaid? مَا لَكُمْ مِنْ وَلِيٍّ ye have no helper (in which case the subject وَلِيٍّ receives the addition of مِنْ

(= French *du*, &c.) as strengthening the negation). In the same way a predicate consisting of a preposition and its noun, or of an adverb, stands before the subject when the latter is undetermined and is not more precisely defined by any qualifying word or phrase, as اِمْرَاَةٌ فِى ٱلدَّارِ in the house is a woman; مِنْهُمْ مَنْ زَعَمَ among them are some who maintain.

NOTE. A predicate of this sort may even stand before a determined subject, but in that case the logical emphasis is on the subject, as عِنْدِى زَيْدٌ, *Zaid* is with me, while in زَيْدٌ عِنْدِى the logical stress is on the predicate: Zaid is with *me*.

142. Verbal adjectives (§ 60 *b*), in virtue of the verbal idea inherent in them, sometimes stand as predicate *before* the noun in the place of a finite verb, as زَيْدٌ ضَارِبٌ أَبُوهُ عَمْرًا Zaid, his father struck Amr = Zaid's father &c. The predicate, thus placed in advance, frequently agrees in gender and number with its subject following, as ٱلْمُؤَلَّفَةِ قُلُوبُهُمْ whose hearts have been inclined (to Islam), but in respect of case it agrees with the word on which this kind of sentence is generally dependent, as بِفَمٍ عَذْبٍ رِيقُهُ with a mouth, whose saliva is sweet; رَأَيْنَا دَوَابَّ مُخْتَلِفَةً أَلْوَانُهَا we found animals, the species of which differed from each other, of different sorts. A circumstantial accusative

(§ 113 b) may also, in this way, refer to a following subject, although it is really dependent on the preceding verbs, as جَاءَ زَيْدٌ رَاكِبًا أَبُوهُ Zaid came, while his father rode.

When the subject of a nominal sentence consists 143. of a demonstrative pronoun, the latter agrees in gender with the following predicate, as هٰذِهِ جَارِيَةٌ this is a female slave.

The predicate of مَا not (often also that of لَيْسَ 144. §§ 50 and 110, and of كَانَ § 110 when occurring with a negative) is <u>introduced by</u> بِ, as مَا هٰذَا بِمَلِكٍ this is no king.

In the relation of subject and predicate (cf. § 119 a) 145. may stand in Arabic:

a. A thing and its dimensions, as اَلْعَمُودُ ثَلْثُونَ ذِرَاعًا the pillar is thirty cubits (high).

b. A thing and that which it resembles, as اَلْبَيْعُ مِثْلُ ٱلرِّبَا selling is the likeness of (is like) usury; and so with كَ (§ 95 f), which likewise may stand in any of the three cases.

c. A thing and its parts, as مُلُوكُ ٱلْفُرْسِ أَرْبَعُ طَبَقَاتٍ the kings of the Persians fall into four divisions.

d. A thing and its material بَعْضُ ٱلْأَصَابِعِ حَدِيدٌ وَبَعْضُهَا خَزَفٌ one part of the toes was of iron and another of clay.

146. In certain cases a pronoun has to be supplied as subject of a nominal sentence, as يُقَالُ لَهُ مُحَمَّدٌ it is said of him "he is Muḥammed", i. e. he is called Muḥammed, prop. = هُوَ مُحَمَّدٌ.

147. The particles إِنَّ (הִנֵּה) behold, and أَنَّ that (cf.
a. § 96 *d*), the compound particles لٰكِنَّ (لَاكِنَّ) nevertheless, كَأَنَّ as if, لِأَنَّ because, and other combinations, and also لَعَلَّ perhaps, لَيْتَ would that, are all followed by a *nominal sentence* the subject of which stands in the *accusative*, as إِنَّ زَيْدًا كَرِيمٌ behold (truly) Z. is generous. The predicate of the nominal sentence following إِنَّ or أَنَّ, if it should consist of an adverb or a preposition with its case (see §§ 139, 141), may stand *before* the subject, which must still be in the accusative, as إِنَّ هُنَا رَجُلًا verily (only in the rarest cases translatable) here is a man; إِنَّ فِي ٱلْقَلْعَةِ سِجْنًا in the citadel is a prison.

NOTE. Sometimes a qualifying phrase consisting of a preposition and its case appears, in addition, before the subject, as إِنَّ لِي إِلَيْكَ حَاجَةً I have a request (to make) of thee.

148. NOMINAL SENT. WITH 'inna AND 'anna.

The corroborative particle لَ (§ 95 g) is frequently b.
prefixed to the predicate after a preceding إِنَّ, as
إِنَّ أَبَانَا لَفِى ضَلَالٍ truly our father is in error; or
even to the subject, as إِنَّ فِى ذٰلِكَ لَعِبْرَةً truly there-
in is an example.

After the particles above mentioned, the pronoun c.
of the 3. pers. sing. masc., as the so-called pronoun
of the fact, is sometimes used as the subject of a
nominal sentence; the predicate, in this case, consists
of a complete sentence (cf. § 139 d), as إِنَّهُ لَا يُفْلِحُ
ٱلْمُجْرِمُونَ of a truth (= the fact is), the evil-doers do
not prosper; قِيلَ أَنَّهُ كَانَ لِمُحَمَّدٍ أَرْبَعُ جَوَارٍ it is related that M. had four female slaves.

While إِنَّ introduces a new and independent sen- 148.
tence, one introduced by أَنَّ always forms part of a.
another sentence, as أَلَمْ تَعْلَمْ أَنَّ ٱللَّهَ عَلَى كُلِّ شَىْءٍ
قَدِيرٌ knowest thou not that God is mighty over all;
here the sentence beginning with أَنَّ is really the object. In لَمْ يُشَكَّ فِى أَنَّهُ أَعْمَى there has never been
any doubt that he is blind, the sentence with أَنَّ is
virtually in the genitive; in بَلَغَنِى أَنَّهُ تَزَوَّجَ it has
reached my ears that he is married, it represents the
subject.

b. Verbal sentences introduced by أَنْ also form in this way an integral part of the principal sentence; a distinction must be made, however, between two varieties of this construction. If the sentence beginning with أَنْ asserts that something is now going on, or that it has now ceased, the verb in the subordinate clause remains in the indicative, as عَجِبْتُ مِنْ أَنْ يَخْرُجَ عَلَيَّ (أَنْ simply or) I am surprised that he takes the field against me, فَفَعَلُوا ذَلِكَ إِلَى أَنْ مَاتُوا and they did this until they died; if, on the other hand, something is conceived as falling in the future and therefore still uncertain, the subjunctive (cf. § 100) is required, as لَكَ أَنْ تَفْعَلَ كَذَا it falls to thee to do so, يَنْبَغِى أَنْ تَحْذَرَ مِنَ ٱلْفَوَاحِشِ it is fit and proper that thou shouldst guard against shameful actions.

NOTE. Sometimes the preposition which indicates the relation of the two parts of the sentence is omitted before أَنْ and أَنَّ, as ذَلِكَ لِأَنَّ = ذَلِكَ أَنَّ this was for the reason that, and it was so, because &c.

c. In the cases discussed in the above sub-section an infinitive may take the place of أَنْ with the finite verb. Quite as frequently as أَنْ in such cases, we find مَا with the finite verb (of course always in the indicative), as عَجِبْتُ مِمَّا ضَرَبْتَ زَيْدًا I am surprised that thou hast struck Zaid = مِنْ ضَرْبِكَ زَيْدًا. The use of this so-called infinitive-*mā* is very common; thus we have it in كَمَا (as)—made up of كَ and مَا—with a verbal sentence: ضُرِبَ زَيْدٌ كَمَا ضُرِبَ عَمْرٌو Zaid was beaten as ʿAmr was beaten.

149. When more than one predicate is required in a nominal sentence, they generally follow each other without a conjunction (cf. §§ 122, 113 *b*, note b), as إِنِّى حَفِيظٌ عَلِيمٌ I am attentive and well-informed. The same is the case with the predicates of the verb كَانَ (which frequently occurs as the substantive verb) and the verbs akin thereto (see § 110), as إِنَّ ٱلْمَمْلَكَةَ تَصِيرُ آخِرَ ٱلْوَقْتِ مُخْتَلِطَةً مُخْتَلِفَةً بَعْضُهَا قَوِىٌّ وَبَعْضُهَا ضَعِيفٌ the kingdom will in the latter days become mixed and a prey to dissension, and one of which one part will be strong and another weak.

150. *a.* In negative verbal sentences we find مَا with the perfect, as مَا شَرِبَ he did not drink, or لَمْ with the apocopated impf. (jussive, cf. § 101 *c*).

b. With the impf. indicative مَا is used, as مَا يَشْرَبُ

he does not drink, or لا with the same tense لَا يَشْرَبُ he does not, or he will not drink.

Other uses of لا are (*a*) with the apoc. impf. (cf. § 101 *b*) and (*b*) with the perfect (cf. § 98 *d*). As negativing an act in the past لا can only stand before the perfect when two perfects come together, as لَا صَدَّقَ وَلَا صَلَّى he neither believed nor prayed, or after sentences with other negatives.

NOTE. A preceding negative, even in the same sentence, is frequently resumed by means of لا, as لَمْ يَجِدِ ٱلْقَرْيَةَ وَلَا صَاحِبَهُ he did not find the village nor yet his friend again.

151.   After the exceptive particle إلّا that which is excepted stands in the *accusative* when a positive sentence precedes, as جَاءَ ٱلنَّاسُ إِلَّا زَيْدًا the people came, except Zaid; when a negative sentence precedes that which is excepted is less frequently in the accusative, but rather, as a rule, in the same case as the word to which the limitation or exception applies, as مَا جَاءَ ٱلْقَوْمُ إِلَّا زَيْدٌ the people came not, except Zaid; مَا مَرَرْتُ بِأَحَدٍ إِلَّا زَيْدٍ I passed no one except Z.; مَا ضَرَبْتُ أَحَدًا إِلَّا عَمْرًا I have struck no one, except 'Amr. Very frequently in such cases it is the exception that brings us the necessary logical complement, as

مَا مَرَرْتُ إِلَّا بِزَيْدٍ I have not passed (anyone) except Zaid, i. e. I have passed only Zaid.

NOTE. Also in the sentence لَا إِلٰهَ إِلَّا ٱللّٰهُ (§ 111) there is no God but Allah, the last word is in the nominative, because it is the logical subject (there is no God, if not Allah; but Allah is). In the sentence لَا حَوْلَ وَلَا قُوَّةَ إِلَّا بِٱللّٰهِ ٱلْعَلِيِّ ٱلْعَظِيمِ there is neither power nor strength except (in union) with Allah, the high and mighty One, the ideas of power and strength (حَوْلٌ وَقُوَّةٌ) must logically be supplied before the exception.

Chapter V. Compound Sentence. (§§ 152—161).

152. Co-ordinate sentences are as a rule joined together by a copulative particle. Thus a simple co-ordinated sentence is usually introduced by وَ (§ 95 i), as دَخَلَ زَيْدٌ وَقَالَ Z. entered and said. فَ (§ 95 e), on the other hand, is used when the connection of the two sentences is less close, when, for example, the second event follows the first only after a certain interval, as مَرِضَ زَيْدٌ فَتُوُفِّيَ Zaid was ill; soon after he died. فَ, accordingly, is often used when the subject is changed, as جَاءَ زَيْدٌ فَقُلْتُ لَهُ Zaid came; and so I said to him. فَإِنَّ with a following nominal sentence expresses the motive of the action and is to be rendered by 'then', 'therefore'.

NOTE a. In lively narrative prose the connective particles are often dispensed with, particularly when the story is told in dialogue form, the words of each speaker being then mostly introduced by a simple قَالَ.

NOTE b. As illustration of the omission of the connectives (asyndeton) must not be quoted certain combinations of two verbs (cf. § 99 note a), in which the second verb denotes rather the end to which some more general activity is directed; such, for example, is the imperfect with verbs denoting a beginning. In other cases, a perfect may be made to depend on a perfect, an imperfect on an imperfect, an imperative on an imperative, as قُمْ اخْطُبْهَا قَامُوا تَقَاتَلُوا they arose and fought with each other; arise and woo her.

NOTE c. Among the connective particles حَتَّى may also, in a certain sense, be reckoned, when it does not introduce a result expected in the future (§ 100), but denotes the actual completion of an action, as in the sentence سَارَ حَتَّى نَزَلَ مَكَّةَ he journeyed until he alighted at Mecca = he journeyed and at last alighted &c. In such cases حَتَّى may also be followed by an imperf. indicative or by أَنْ with a nominal sentence.

153.   *Relative sentences* or clauses are of two kinds, those which do not accompany a noun and those which do accompany and qualify a noun. As regards the asyndetical connection of several qualifications, the latter class is subject to the same treatment as the qualifying adjuncts discussed in §§ 120—122.

154.   Those relative sentences that do not depend on or qualify a noun are introduced either by اَلَّذِى (see § 14 a) he that, that which, whoso, &c., which is

declinable and always determined, or by the indeclinable pronouns مَنْ (he that, one that, whosoever, those that, such...as) and مَا (that which, a thing that, what). The former is sometimes determined, sometimes undetermined. Exx: اَلَّذِينَ كَفَرُوا بِآيَاتِنَا هُمْ أَصْحَابُ ٱلْمَشْأَمَةِ those that reject our revelations, they will be the people of the left hand (اَلَّذِينَ is here in the nom. as being the subject); أَأَسْجُدُ لِمَنْ خَلَقْتَ طِينًا (the devil said:) Shall I fall down before one whom thou hast formed of clay (مَنْ is here in the genit.)? يَقُولُونَ بِأَفْوَاهِهِمْ مَا لَيْسَ بِقُلُوبِهِمْ they speak with their mouth what is not in their hearts (مَا is here accus.).

A relative clause is made to follow and qualify **155.** a substantive by means of اَلَّذِى only when the substantive in question (the antecedent) is *determined*; with it اَلَّذِى agrees in gender and number, as ضَرَبْتُ ٱلرَّجُلَ ٱلَّذِى جَاءَ I struck the man that came. The explanation of this is that اَلَّذِى is originally not a relative in our sense of that word, but a demonstrative, and as such it is always determined. The above sentence, for example, means, strictly speaking: I struck that man there, he came. On the other hand the relative clause is appended *without* اَلَّذِى when the antecedent is undetermined, as ضَرَبْتُ رَجُلًا جَآءَ I struck a man who came (prop. I struck a man, he came).

NOTE. اَلَّذِى is also dispensed with when the antecedent is only determined in a general sense (i. e. when it has the generic article see § 118c), as كَمَثَلِ ٱلْحِمَارِ يَحْمِلُ أَسْفَارًا like *an* ass that carries books.

156. The relative clause, which we have seen to be strictly speaking merely a verbal or a nominal sentence subordinated to an antecedent noun, ought by rule to contain a pronoun referring back to this antecedent, as اَلرَّجُلُ ٱلَّذِى أَبُوهُ غَنِىٌّ the man whose father is rich; كَانَ لَهُ ٱبْنٌ سُمِّىَ مُحَمَّدًا he had a son, who was named M. (in this case the pronoun is implied in the verb); رَجُلٌ يُقَالُ لَهُ زَيْدٌ a man who is named Z. (prop. of whom it is said: [he is] Zaid, cf. § 146). The pronoun which in this way points back to the antecedent may stand in any part of the relative sentence; thus in the sentence قَدْ قَرُبَ إِلَيْهِ ٱلْجَيْشُ ٱلَّذِى ظَنَّ أَنَّهُ بَعِيدٌ the army had come up close to him, regarding which he thought that it was still at a distance, it does not appear till we reach the sentence which is subordinated by أَنْ to the verb ظَنَّ.

157. CIRCUMSTANTIAL CLAUSES.

Collectives which denote living creatures (cf. § 136 d) may be followed here also by a plural verb, as قَوْمٌ يُؤْمِنُونَ people that believe.

NOTE a. The omission of the pronoun, however, is not unfrequent, especially when it would merely consist of a suffix of the 3. person, as نَدِمْتُ عَلَى مَا قُلْتُ for قُلْتُهُ I regret what I said.

NOTE b. In certain cases the antecedent may be repeated in the relative clause; indeed, this is the favourite construction with كُلّ as كَانَ عِنْدَ هُبَلَ قِدَاحٌ سَبْعَةٌ كُلُّ قِدْحٍ مِنْهَا فِيهِ كِتَابٌ the (idol) Hubal had seven arrows (for casting the lot), of which each single arrow had writing upon it.

157. A special kind of subordinate sentence is the *circumstantial clause*. Such a clause may consist:

1) Of a *nominal sentence* introduced by the particle *a.* وَ, the subject of which may have been already mentioned or may be something quite new, as مَاتَتْ آمِنَةُ وَهِيَ رَاجِعَةٌ إِلَى مَكَّةَ Amina died while she was returning to Mecca; مَاتَ زَيْدٌ وَٱبْنُهُ صَغِيرٌ Zaid died while his son was still young; with a compound nominal sentence سَارَ وَهُوَ يَقْصِدُ ٱلْمَدِينَةَ he journeyed taking Medina as his goal. A sentence, whose predicate consisting of a preposition and its case comes before its subject, acc. to § 141, may stand as a circumstantial clause, *without* وَ, as خَرَجْتُ (وَ)فِي يَدِي قَوْسٌ I went out with a bow in my hand.

9*

b. 2) Of a *verbal sentence* frequently; in this case the imperf. either stands alone or is preceded by وَقَدْ. When the sentence is a negative one, the negative is لَا or وَمَا; or the verb may stand in the apoc. impf. with لَمْ or وَلَمْ (as the negation of the perf.). We may also have the perfect with وَقَدْ or وَكَانَ, when negative with وَمَا; thus we get the following: جَآءَ زَيْدٌ يَضْحَكُ Z. came laughing; قَالَ أَنَّى يَكُونُ لِي غُلَامٌ وَكَانَتِ امْرَأَتِي عَاقِرًا وَقَدْ بَلَغْتُ مِنَ ٱلْكِبَرِ عِتِيًّا (Zakarīya) said: how shall I have a male child, seeing my wife is barren and I have reached too great an age; دَخَلَ ٱلْبَيْتَ لَا يُسَلِّمُ عَلَىَّ he entered the room without greeting me.

NOTE. In contrast to the stiffer accusative of condition (§ 113 *b*) the verbal circumstantial clause expresses the commencement of the action; there is very little difference, however, between جَآءَ زَيْدٌ ضَاحِكًا and جَآءَ زَيْدٌ يَضْحَكُ.

158. In *temporal* clauses (also in conditional clauses)
a. which are formed with the particle إِذَا when, if, we find in the protasis as well as in the apodosis the perfect in the sense of our present or future, إِذَا رَاضَ يَحْيَى ٱلْأَمْرَ ذَلَّتْ صِعَابُهُ when John takes the thing in hand, its difficulties are easily surmounted.

NOTE a. The imperfect may also stand after إِذَا if the action takes place repeatedly. Should إِذَا be followed by a compound nominal sentence, as إِذَا ٱلْجَحِيمُ سُعِّرَتْ when hell is heated, it is considered that this is but another way of writing what we should expect to find expressed in a verbal sentence (and so with إِنْ).

NOTE b. A sentence with إِذَا may also be inserted between two closely related words, or rather it is to be regarded as forming with its apodosis a complete unity. Thus: فَإِنَّهُمَا بَابَانِ إِذَا فُتِحَا لَمْ يُغْلَقَا there were two gates, which when they were opened could not be shut (again). In the apodosis to إِذَا a perfect is found where we should expect an imperfect (cf. § 99c), as كَانُوا إِذَا أَسَرُوا رَجُلًا وَأَطْلَقُوا جَزُّوا نَاصِيَتَهُ they were wont, when they captured a man and then released him, to cut off his front lock of hair. Very frequently a sentence like this, with إِذَا, is inserted between حَتَّى (§ 152 note c) and its proper verb, as تَبِعْتُهُ حَتَّى إِذَا دَخَلَ ٱلدَّارَ أَدْرَكْتُهُ I followed him until I overtook him as he entered the house.

مَا in the sense of 'so long as' takes the perfect, as مَا دُمْتُ أَنَا شَاكِرٌ so long as I live I shall be thankful.

159. In sentences containing the notion of a *condition* which is the case after إِنْ if, مَنْ if anybody, مَا if anything, مَهْمَا whatsoever, كَيْفَ, كَيْفَ مَا how, howsoever, مَتَى when &c. the perf. is used in the sense of our present or future, and so too in the apodosis, as إِنْ فَعَلْتَ ذَلِكَ هَلَكْتَ if thou doest that, thou

wilt perish; مَنْ جَالَ نَالَ whoso seeketh, findeth (if any one seeks, he finds).

> NOTE. If the perf. is meant to retain its proper force in the protasis, the verb كَانَ is placed after إِنْ, as إِنْ كَانَ قَمِيصُهُ قُدَّ مِنْ قُبُلٍ فَصَدَقَتْ if his camisole is torn in front, she has told the truth.

160. The particles above mentioned may also take the

a. apoc. impf. in protasis and apodosis alike, as إِنْ تَصْبِرُوا يُمَدِّدْكُمْ رَبُّكُمْ if ye wait patiently, God will help you.

b. The apoc. impf. also stands in the apodosis after an imperative (with conditional force) in the protasis, as عِشْ قَنِعًا تَكُنْ مَلِكًا live contentedly (i. e. if thou live &c.) thou wilt be a king.

c. An apoc. impf. in the protasis may be followed by a perfect in the apodosis, as إِنْ تَصْبِرْ ظَفِرْتَ if thou wait patiently, thou wilt gain the victory. If the clauses are both negative, we have لَمْ with the apoc. impf., as إِنْ لَمْ يَبْرَحْ لَمْ أَرْضَ if he does not go away, I am not satisfied.

> NOTE. Occasionally the apodosis of a conditional sentence is wanting, e. g. إِنْ كَانَ هٰذَا if this is so—supply: then it is well (Arab. فَبِهَا).

161. Before the apodoses of conditional sentences, other than those discussed in § 159—160 we find the particle فَ, which is employed:

161. CONDITIONAL CLAUSES.

1) When the apodosis is a nominal sentence, as إِنْ عَصَى فَوَيْلٌ لَهُ if he is refractory, then alas for him! Also before sentences with إِنَّ and before interrogative sentences.

2) When the apodosis is a verbal sentence, of which the perfect is intended to retain its force as a perfect (cf. § 159 note), especially, too, when قَدْ (cf. § 98 e) is employed, as إِنْ أَسْلَمُوا فَقَدِ ٱهْتَدَوْا وَإِنْ تَوَلَّوْا فَإِنَّمَا عَلَيْكَ ٱلْبَلَاغُ if they become Moslems, then have they come to the right way, and if they turn aside, then thou hast but to announce the message.

3) When the apodosis is a verbal sentence that contains an impf. with one of the particles لَنْ, سَ, سَوْفَ, or that expresses a command or a wish, as إِنْ كُنْتَ فِي قَوْمٍ فَٱحْلُبْ فِي إِنَائِهِمْ if thou findst thyself among people, milk into their pail.

APPENDIX.

COMPUTATION OF TIME.

a. *Names of the Days of the Week.*

In the following list the various names may also be used with the word for day, يَوْمٌ omitted.

1. يَوْمُ ٱلْأَحَدِ (1st day) Sunday.
2. يَوْمُ ٱلْاِثْنَيْنِ (2nd day) Monday.
3. يَوْمُ ٱلثَّلَاثَآءِ (3rd day) Tuesday.
4. يَوْمُ ٱلْأَرْبِعَآءِ (4th day) Wednesday.
5. يَوْمُ ٱلْخَمِيسِ (5th day) Thursday.
6. يَوْمُ ٱلْجُمْعَةِ (day of assembly) Friday.
7. يَوْمُ ٱلسَّبْتِ (Sabbath) Saturday.

b. *Names of the Months.*

In the names of the months the word شَهْرُ, month, may be prefixed in the constr. state throughout; indeed, as the following table shows, some of the names are always so written.

1. اَلْمُحَرَّم al-Muḥarram.

2. صَفَر Ṣafar.

3. شَهْرُ رَبِيعٍ ٱلْأَوَّلِ the first Rabī'.

4. شَهْرُ رَبِيعٍ ٱلثَّانِى the second Rabī'.

5. جُمَادَى ٱلْأُولَى the first Ǧumādā.

6. جُمَادَى ٱلْآخِرَة the latter Ǧumādā.

7. رَجَب Raǧab.

8. شَعْبَان Ša'bān.

9. رَمَضَان Ramaḍān (the month of fasting).

10. شَوَّال Šawwāl.

11. ذُو ٱلْقَعْدَة Ḏu-lḳa'da.

12. ذُو ٱلْحِجَّة Ḏu-lḥiǧǧa (month of the pilgrimage, ḥaǧǧ).

c. *The Year.*

The Moslems reckon by lunar years of 354 days; their first year is usually considered as beginning at the date of the Christian era given below. In calculating from one era to the other, it may be reckoned that 33 solar years are equal to 34 lunar years.

In the works of European scholars it is customary, by means of comparative tables, to give the precise day of our era with which each Moslem year begins (see the Bibliography). The following short table will be useful in helping to a rapid approximation of the date required.

The Moslem year			1	began	16. July	622	A.	D.
„	„	„	101	„	24. July	719	„	„
„	„	„	201	„	30. July	816	„	„
„	„	„	301	„	7. Aug.	913	„	„
„	„	„	401	„	15. Aug.	1010	„	„
„	„	„	501	„	22. Aug.	1107	„	„
„	„	„	601	„	29. Aug.	1204	„	„
„	„	„	701	„	6. Sept.	1301	„	„
„	„	„	801	„	13. Sept.	1398	„	„
„	„	„	901	„	21. Sept.	1495	„	„
„	„	„	1001	„	8. Oct.	1592	„	„
„	„	„	1101	„	15. Oct.	1689	„	„
„	„	„	1201	„	24. Oct.	1786	„	„
„	„	„	1301	„	2. Nov.	1883	„	„
„	„	„	1313	„	24. June	1895	„	„

LITERATURE.

A history of Arabic literature as a whole, or even of particular parts of it, does not exist, for the work of Hammer-Purgstall (Litteraturgeschichte der Araber, von ihrem Beginn bis zu Ende des zwölften Jahrhunderts der Hidschret. 7 Bände. Wien 1850—56. 4⁰.) must be described as premature and as useless by reason of its numerous mistakes. An acquaintance with Arabic literature must therefore be got partly from works by Arabs on the history of their literature, partly from European catalogues. In the course of the present century numerous works, including not a few specimens of the earlier literature, have been printed in the East, especially in Cairo (government press in Būlāk), Beirūt (where there is an excellent press managed by the Jesuits) and Constantinople; also in Persia, India and the island of Java. We must, in particular, mention the great quantity of valuable Arabic manuscripts that still await publication both in European and eastern libraries. A synopsis of such catalogues of these MSS. as have hitherto appeared will be found below.

In the following selection, books of special importance are marked with a star, those recommended to beginners with a dagger.

A. BIBLIOGRAPHY.

I. Printed Works.

a Written by Orientals.

*Kitāb al-Fihrist (by *Ibn abī Ya'ḳūb an-nadīm*; wrote in the year 377 H., beg. 3. May 987) mit Anmerkungen herausgegeben von *Gustav Flügel*. Nach dessen Tode besorgt von *Johannes Rödiger* und *August Müller*. 2 voll. Leipzig 1871—2.

*Lexicon bibliographicum et encyclopaedicum a Mustapha ben Abdallah Katib Jelibi dicto et nomine *Haji Khalfa* (*Ḥaǧǧi Ḥalīfa* † 1658) celebrato compositum. Ad codicum Vindobonesium Parisiensium et Berolinensis fidem primum edidit latine vertit et commentario indicibusque instruxit *Gustavus Flügel*. Leipzig-London 1835—1858. 7 voll. 4⁰.

β *Written by Europeans*.

Bibliotheca arabica. Auctam nunc atque integram edidit *D. Christianus Fridericus de Schnurrer*. Halae ad Salam 1811.
†Bibliotheca orientalis. Manuel de Bibliographie orientale. I. contenant les livres arabes, persans et turcs imprimés depuis l'invention de l'imprimerie jusqu'à nos jours tant en Europe qu'en Orient etc. par *J. Th. Zenker*. Leipzig 1846. — Bibliotheca orientalis. Manuel de Bibliographie orientale. II. contenant 1. supplément du premier volume. 2. Littérature de l'Orient chrétien. 3. Littérature de l'Inde etc. Par *J. Th. Zenker*. Leipzig 1861.
†(Euting) Katalog der kaiserlichen Universitäts- und Landesbibliothek in Strassburg. Arabische Litteratur. Strassburg 1877. 4º.
Bibliographie des ouvrages arabes ou relatifs aux Arabes publiés dans l'Europe chrétienne de 1810 à 1885 par *Victor Chauvin*. I. Préface. — Table de Schnurrer. — Les Proverbes. Liége 1892 (is being continued).
Wissenschaftlicher Jahresbericht über die morgenländischen Studien, von 1844 an in Zeitschrift der Deutschen Morgenländischen Gesellschaft Leipzig 1847 ff. The annual reports on works published up to 1858 appeared in the *Zeitschrift*, those for the years 1859—61, 62—67 (one part), autumn 1877—81 appeared as independent publications.
Bibliotheca orientalis oder eine vollständige Liste der im Jahre 1876 in Deutschland, Frankreich, England und den Colonien erschienenen Bücher, Broschüren, Zeitschriften, u. s. w. über die Sprachen, Religionen, Antiquitäten, Literaturen, Geschichte und Geographie des Ostens, zusammengestellt von *Karl Friederici*. Leipzig. 8 years (to 1883).
Bibliography for 1883—85 (not completed) in the Literatur-Blatt für orientalische Philologie unter Mitwirkung von Dr. Johannes Klatt herausgegeben von Prof. Dr. *Ernst Kuhn*. 1883—85.
*Orientalische Bibliographie . . . herausgegeben von *A. Müller*, now *E. Kuhn*. Berlin 1888 ff.
Katalog der Bibliothek der Deutschen Morgenländischen Gesellschaft. I. Druckschriften und Ähnliches. Leipzig 1880 (a new and largely augmented edition will appear in a year or two).
A. G. Ellis, Catalogue of the Arabic books in the British Museum Vol I. A-L. London 1894.
For works from oriental presses an important guide is: *E. J. Brill*, Catalogue périodique de livres orientaux I—IX, Leide 1883 ff. (To parts I—VII Index de noms d'auteurs et de noms de livres, ib. 1889).

II. Manuscripts.

(Die Handschriftenverzeichnisse der königlichen Bibliothek in Berlin. Vols. 7 ff.). Verzeichniss der arabischen Handschriften von *W. Ahlwardt*. 4⁰. 1. Band. Berlin 1887; 2. Bd. 1889; 3. Bd. 1891; 4. Bd. 1892; 5. Bd. 1893; 6. Bd. 1894. A 7th and last vol. will appear soon.

(*Halle*) Katalog der Bibliothek der Deutschen Morgenländischen Gesellschaft. II. Handschriften u. s. w. Leipzig 1881.

Verzeichnis der orientalischen Handschriften der Bibliothek des *Halle*'schen Waisenhauses von *Fr. Aug. Arnold* und *August Müller*. (Programm der Lateinischen Hauptschule). Halle 1876. 4⁰.

(University Library, *Leipzig*) Die Refaïya. Von Prof. *Fleischer:* Zeitschrift der Deutschen Morgenländischen Gesellschaft. 8, S. 573—584.

(Municipal Library in *Leipzig*) Catalogus librorum manuscriptorum, qui in bibliotheca senatoria civitatis Lipsiensis asservantur, ed. Naumann. Codices orientalium linguarum descripserunt *H. O. Fleischer* et *Fr. Delitzsch*. Grimmae 1838. 4⁰.

Catalogus codicum manuscriptorum orientalium Bibliothecae regiae *Dresdensis*. Scripsit et indicibus instruxit *H. O. Fleischer*. Lipsiae 1831. 4⁰.

Die arabischen Handschriften der herzoglichen Bibliothek zu *Gotha*. Verzeichnet von *Wilhelm Pertsch*. 5 Bände. Gotha 1878—1892. (Also w. the title: Die orientalischen Handschriften der h. B. zu G. Dritter Theil).

Die arabischen Handschriften der K. Hof- und Staatsbibliothek in *München*, beschrieben von *Joseph Aumer*. München 1866. (Catalogus codicum manuscriptorum Bibliothecae regiae Monacensis. Tomi primi pars secunda.)

(*Tübingen* University Library) Catalog arabischer Handschriften in Damaskus gesammelt von *J. G. Wetzstein*. Berlin 1863.

Catalogus librorum manuscriptorum orientalium in bibliotheca academica *Bonnensi* servatorum adornavit *Joannes Gildemeister*. Bonnae 1864—1876. 4⁰.

Katalog der hebräischen, arabischen, persischen und türkischen Handschriften der kaiserlichen Universitäts- und Landesbibliothek zu *Strassburg*. Bearbeitet von *S. Landauer*. Strassburg 1881. 4⁰.

Die arabischen, persischen und türkischen Handschriften der kaiserlich-königlichen Hofbibliothek zu *Wien*. Von *Gustav Flügel*. 3 Bände. Wien 1865—7. 4⁰.

(*Copenhagen*) Codices orientales Bibliothecae regiae Havniensis enumerati et descripti a *N. L. Westergaard* etc. II. Codices hebr. et arab. Hafniae 1851.

Codices Orientales bibliothecae regiae universitatis *Lundensis* recensuit *Carolus Johannes Tornberg*. Lundae 1850.

Codices Arabici, Persici et Turcici bibliothecae regiae universitatis *Upsaliensis*. Disposuit et descripsit *C. T. Tornberg*. Upsaliae 1849. 4⁰.

(*Paris*) Catalogue des manuscrits arabes de la Bibliothèque Nationale par le Baron *de Slane*. Pr. Fascicule. Paris 1883. Sec. Fasc. 1889. Trois. Fasc. 1895. 4⁰. (To be continued.)

Catalogue général des manuscrits des bibliothèques publiques de France. Départements. Tome VI (p. 437—482). Marseille. Par M. l'abbé *Albanès*. Paris 1892. — Tome XVIII. Alger. Par *E. Faynan*. Paris 1893.

(*Leide*) Catalogus codicum orientalium Bibliothecae academiae *Lugduno Batavae* I. II. auctore R. P. A. *Dozy*. III. IV. auct. *P. de Jong* et *M. J. de Goeje*. V. auctore *M. J. de Goeje*. VI. auctore *M. Th. Houtsma*. Lugduni Bavatorum 1851—77. — Editio secunda. Vol. I auctoribus *M. J. de Goeje* et *M. Th. Houtsma*. Lugduni Bat. 1888.

(*London*) Catalogus codicum manuscriptorum orientalium qui in *Museo Britannico* asservantur. Pars secunda codices arabicos amplectens. Londini 1846. fol.

(*London*) Supplement to the Catalogue of the Arabic manuscripts in the British Museum (By Charles Rieu). London 1894, 4⁰.

London) A catalogue of the Arabic manuscripts in the library of the *India Office*. By *Otto Loth*. London 1877. 4⁰.

(*Oxford*) Bibliothecae Bodleianae codicum manuscriptorum orientalium, videlicet hebraicorum, chaldaicorum, syriacorum, aethiopicorum, arabicorum, persicorum, turcicorum, copticorumque catalogus a *Joanne Uri* confectus. Pars Prima Oxonii 1787. — Partis secundae volumen primum arabicos complectens confecit *Alexander Nicoll*. Oxonii 1821. fol.

(*Cambridge*) Catalogus Bibliothecae Burckhardtianae cum appendice librorum aliorum orientalium in Bibliotheca Academica *Cantabrigensitis* asservatorum — confecit *T. Preston*. Cantabrigiae 1853. 4⁰.

Catalogue of the Oriental Manuscripts in the Library of King's College, *Cambridge*. By *Edward Henry Palmer*: Journal of the Roy. As. Society of Gr. Britain and Ireland. New Series III. 105 ff.

A descriptive Catalogue of the Arabic, Persian and Turkish Manuscripts in the Library of Trinity College, *Cambridge*. By *E. H. Palmer*. Cambridge and London 1870.

(*Escurial*) Bibliotheca arabico-hispana *Escurialensi* sive Librorum omnium Mss. quos Arabice ab auctoribus magnam partem Arabo-Hispanis compositos Bibliotheca Coenobii Escurialensis complectitur recensio et explanatio operâ et studio *Michaelis Casiri* etc. 2 tomi. Matriti 1760. fol. — Les manuscrits arabes de l'Escurial décrits par *Hartwig Dérenbourg*. Tome premier. Paris 1884.

Catálogo de los Manuscritos árabes existentes en la Biblioteca Nacional de *Madrid* (*F. G. Roblès*). Madrid 1889.

(*Florence*) Bibliothecae Mediceae Laurentianae et Palatinae Codicum manuscriptorum orientalium catalogus, *Steph. Evod. Assemanus* recensuit. Florentiae 1742. fol.

(*Venice*) Catalogo dei Codici manoscritti orientali della Biblioteca Naniana, compilato dell' abbate *Simone Assemani*. 2 Part. Padova 1787—1792. 4⁰.

Remarques sur les manuscrits orientaux de la Collection Marsigli a *Bologne* suivies de la liste complète des Manuscrits arabes de la même collection par le Baron *Victor Rosen*. Roma 1885 (atti della R. Academia dei Lincei. Serie 3ª. Vol. XII).

Milan) Catalogo dei Codici arabi, persiani e turchi della Biblioteca Ambrosiana (*Hammer-Purgstall*): Biblioteca Italiana t. XCIV, pp. 22 and 322.

Cataloghi dei codici orientali di alcune biblioteche d'Italia. 5 fasc. Firenze 1878—1892.

Catalogue des manuscripts et xylographes orientaux de la Bibliothèque Impériale publique de *St. Pétersbourg*. St. Pétersbourg 1852.

(*St. Petersburg*) B. *Dorn*, Catalogue des ouvrages arabes, persans et turcs, publiés à Constantinople, en Egypte et en Perse, qui se trouvent au Musée asiatique de l'Académie. — Chronologisches Verzeichniss der seit dem Jahre 1801 bis 1866 in Kasan gedruckten arabischen, türkischen, tatarischen und persischen Werke, als Katalog der in dem asiatischen Museum befindlichen Schriften: Mélanges asiatiques tirés du Bulletin de l'Académie Impériale des sciences de St. Pétersbourg. Tome V. Livr. 5. St. Pétersbourg 1867.

(*St. Petersburg*) Notices sommaires des manuscrits arabes du Musée asiatique par le Baron *Victor Rosen*. St. Pétersbourg 1881.

(*St. Petersburg*) Les manuscrits arabes de l'Institut des langues orientales décrits par le Baron *Victor Rosen*. St. Pétersbourg 1877.

(*J. M. E. Gottwald*) description of the Arabic Manuscripts in the Library of the Imperial University of *Kasan*. Kasan (no date) [1885]. In Russian.

(*Cairo*) Fihrist al-kutub al-ʿarabīya al-maḥfūza bil-kutubḫāna al-ḫedīwīye el-kā'ine biserāi derb al-gamāmīz. (Under the management of *Spitta* and *Vollers*.) 7 vols. Cairo 1301—1308. Second Edition. Vol. I 1310.

Catalog der mektebe ʿumūmīye in *Damascus*. Damascus 1299. 4⁰.

Studia Sinaitica No. III. Catalogue of the Arabic Mss. in the Convent of S. Catharine on Mount *Sinai* compiled by *Margaret Dunlop Gibson*. London 1894.

(*Batavia*) *Friedrich*, Codicum arabicorum in Bibliotheca Societatis Artium et Scientiarum quae *Bataviae* floret asservatorum Catalogus. Absolvit indicibusque instruxit *L. W. C. van den Berg*. Bataviae et Hagae 1873.

B. INTRODUCTION.

General.

Borhân-ed-dini es-*Sernûdji* (as-Sarnūǵī lived at the and of the 12th century of our era) Enchiridion studiosi. Arabice edidit latine vertit et lexico explanavit *Carolus Caspari*. Praefatus est *H. O. Fleischer*. Lipsiae 1838. 4⁰.
Einleitung in das Studium der Arabischen Sprache bis Mohammed und zum Theil später . . . von *G. W. Freytag*. Bonn 1861.
Orientalische Skizzen. Von *Theodor Nöldeke*. Berlin 1892. Translated, with the title 'Sketches form Eastern History' by *J. S. Black*. London and Edinburgh 1892.
De auctorum graecorum versionibus et commentariis syriacis, arabicis, armeniacis persicisque commentatio quam scripsit *Joannes Georgius Wenrich*. Lipsiae 1842. 1845.

C. CHRESTOMATHIES.

*†*R. Brünnow*, Chrestomathy of Arabic Prose-Pieces. Berlin and London 1895.
†Chrestomatia arabica quam e libris Mss. vel impressis rarioribus collectam edidit *Fr. A. Arnold*. Pars I. Textum continens. Pars II. Glossarium continens. Halis 1853.
†Chrestomathie Arabe, ou extraits de divers écrivains Arabes, tant en prose qu'en vers à l'usage des élèves de l'école spéciale des langues orientales vivantes; par *A. J. Sylvestre de Sacy*. II. éd. corr. et augm. Paris 1826. 3 vol.; Tome IV Anthologie grammaticale arabe. Paris 1829.
†Chrestomathie élémentaire de l'Arabe littéral avec un glossaire par *H. Dérenbourg* et *J. Spiro*. 2 ed. Paris 1892.
Joh. Godofr. Lud. Kosegartenii Chrestomathia arabica ex codicibus manuscriptis Paris. Goth. et Berol. collecta atque tum adscriptis vocalibus, cum additis lexico et adnotationibus explanata. Lipsiae 1828.
Georg. Guil. Freytag, Chrestomathia arabica, grammatica historica in usum scholarum Arabicarum ex codd. ineditis conscripta. 8⁰ maj. Bonnae 1834.

† Thier und Mensch vor dem König der Genien. Ein arabisches Mährchen aus den Schriften der lauteren Brüder in Basra im Urtext herausgegeben von *Fr. Dieterici*. 2. Ausgabe." Leipzig 1881. — Arabisch-deutsches Wörterbuch zum Koran und Thier und Mensch von *Fr. Dieterici*. 2. Aufl. Leipzig 1894.
Brevis chrestomathia arabica. In usum scholarum ed. *Joh. Bollig*. Roma 1881.
Chrestomatia arábigo-española por *Fr. J. Lerchundi y Fr. J. Simonet*. Granada 1881.
Girgas and *de Rosen*. Arabic Chrestomathy (in Russian). St. Petersburg 1875. 1876. — Dictionary to the Chrestomathy and to the Koran by *W. Girgas*. Kasan 1881 (in Russian).
An Arabic reading-book compiled by *W. Wright*. Part first, The texts. London 1870.
Maġānī el-adab fī ḥadāik el-'arab. 6. Ed. Beirut 1885 ff. Jesuit Press. 6 vols. Šarḥ maġānī el-adab (Notes &c.). 4 vols. ib. 1886—8.

D. GRAMMARS &c.

a Written by Orientals.

*al-*Muzhir* fī 'ulūm el-luġa, philological Encyclopaedia by Ġalāl ad-dīn as-*Suyūṭī* († 911 H., beg. 4. June 1505, cf. for as-Sujūṭī *Goldziher* in den Sitzungsber. d. kais. Akademie der Wiss. zu Wien. Phil.-histor. Cl. LXIX. Bd. 1. S. 7 ff.) Bulak 1282.

*Le livre de Sībawaihī, traité de grammaire arabe par Siboûya, dit *Sībawaihī* († 180 H., beg. 16. March 796). Texte arabe publié d'après les manuscrits du Caire, de l'Escurial, d'Oxford, de Paris, de St. Pétersbourg et de Vienne par *Hartwig Derenbourg*. Tome I, Paris 1881. Tome II, Paris 1889. — *Sībawaihī's* Buch über die Grammatik nach der Ausgabe von H. Derenbourg und dem Commentar des Sirâfî übersetzt und erklärt . . . von *G. Jahn*. 1.—8. Lieferung. Berlin 1894. 1895.

*Al-Mufaṣṣal, opus de re grammatica arabicum auctore Abu 'l-Kāsim Maḥmūd bin 'Omar Zamaḫšario (*az-Zamaḫšari* † 538 H., beg. 16. July 1143) ed. *J. P. Broch*. Editio altera. Christianiae 1879. — Also: *Ibn Jaʿīš* († 643 H., beg. 29. May 1245) Commentar zu Zamachšarī's Mufaṣṣal. Nach den Handschriften . . . . . herausgeg. u. s. w. von Dr. *G. Jahn*. Erster Band. Leipzig, 1882. Zweiter Band. Leipzig 1886. 4⁰.

*Alfijjah, Carmen didacticum grammaticum auctore Ibn Mālik († 672 H., beg. 18. July 1273) et in Alfijjam commentarius quem conscripsit Ibn Akil (*Ibn ʿAḳīl* † 769 H., beg. 28. Aug. 1367) ed. *Fr. Dieterici*. Lipsiae 1851. — Ibn ʿAḳīl's Commentar zur Alfijja des Ibn Mālik aus dem Arabischen zum ersten male übersetzt von *Fr. Dieterici*. Berlin 1852.

al-Aǧurrūmiyya, Arabic Grammar by *Ibn Aǧurrūm* aṣ-Ṣinhāǧī († 723 H., beg. 10. January 1323). Often printed with and without Commentaries. Cf. *E. Trumpp*, Einleitung in das Studium der arabischen Grammatiken. Die Ajrummiyyah des Muhammad bin Daud. München 1876. On this work see *Fleischer* in Zeitschrift der D. Morgenl. Ges. 30 (1876), pp. 487—513; reprinted in Kleinere Schriften II (Leipzig 1888), pp. 75—106. Text also printed in Brünnow's Chrestomathy.

Kāfiya fin-naḥū, Syntax by *Ibn al-Ḥāǧib* († 646 H., beg. 26. April 1248). Frequently printed in the East.

Muġni al-labīb, Grammar composed by *Ibn Hišām* al-Anṣārī († 762 H., beg. 11. Nov. 1360). Another grammatical work by the same author bears the title: Kaṭar an-nadā wa-ball aṣ-ṣadā; a third Šuḏūr aḏ-ḏahab. All three works have been frequently printed in the East.

al-Ḥarīrī's († 516 H., beg. 16. July 1143) Durrat al-ġawwāṣ, herausgegeben von *Heinrich Thorbecke*. Leipzig 1871. (On errors of speech). With the commentary of al-Ḥafāǧī, Constantinople 1299. Cf. Le livre des locutions vicieuses de Djawālīkī publié par *Hartwig Derenbourg* (al-Ǧawālīkī † 465 H., beg. 17. Sept. 1072) in Morgenländische Forschungen. Leipzig 1875.

Ṭarīka mustaḥdata fī tashīl al-ḫaṭṭ al-ʿarabī. Calligraphic models 12 parts. Beirut 1891.

β *Written by Europeans.*

*Die grammatischen Schulen der Araber nach den Quellen bearbeitet von *G. Flügel*. Erste Abthl. Leipzig 1862. Abhandlungen der Deutschen Morgenl. Ges. II. Band. Nr 4. (This work gives a list of grammarians to about the year 1000 of our era).

† Dr. *C. P. Caspari's* Arabische Grammatik. Fünfte Auflage bearbeitet von *August Müller*. Halle 1887. — Grammaire arabe de C. P. Caspari traduite de la quatrième édition allemande et en partie remaniée par *E. Uricoechea*. Bruxelles 1880. — A Grammar of the Arabic Language translated from the German of Caspari and edited, with numerous additions and corrections by *W. Wright*. 2. ed. 2 vol. London 1874—5. A 3rd edit. is announced.

Geo. Henric. Aug. Ewald. Grammatica critica linguae arabicae cum brevi metrorum doctrina. Lipsiae 1831—1833. II vol.

*Grammaire arabe à l'usage des élèves de l'école spéciale des langues orientales vivantes; avec figures. Par M. le Bon *Silvestre de Sacy*. Seconde édition, corrigée et augmentée, à laquelle on a joint un traité de la prosodie et de la métrique des Arabes. 2 tom. Paris 1831. — Very important notes and corrections will be found in

Fleischer, „Beiträge zur arabischen Sprachkunde": Berichte über die Verhandlungen der kgl. sächsischen Gesellschaft der Wissenschaften zu Leipzig. Philologisch-historische Classe. 1863 (p. 93 ff.); 1864 (p. 265 ff.); 1866 (p. 286 ff.); 1870 (p. 227 ff.); 1874 (p. 71 ff.); 1876 (p. 44 ff.); 1878 (p. 64 ff.); 1880 (p. 89 ff.); 1881 (p. 117 ff.); 1883 (p. 72 ff.); 1884 (p. 272 ff.); conf. 1856 (p. 1 ff.); 1862 (p. 10 ff.) Reprinted in Kleinere Schriften von Dr. *H. L. Fleischer*, vol. I, 1st. and 2nd. parts, Leipzig 1886; the two last articles in vol. II, part 1. Leipzig 1888.

J. G. L. Kosegarten. Grammatica linguae arabicae pp. 1—688, without title and date, incomplete. (Very rare).

Mortimer Sloper Howell. A Grammar of the Classical Arabic Language, translated and compiled from the Works of the most Approved Native or Naturalized Authorities. Published under the Authority of the Government of the N.-W. Provinces. In an Introduction and Four Parts. 3 vols. Allahabad 1880. 1883. 1886.

Grammaire arabe composée d'après les sources primitives par le P. *Donat Vernier*, S. J. Tome I. Beyrouth 1891; Tome II. 1892.

Darstellung der arabischen Verskunst mit sechs Anhängen u. s. w. nach handschriftlichen Quellen bearbeitet und mit Registern versehen von *G. W. Freytag*. Bonn 1830.

Théorie nouvelle de la métrique arabe précédée de considérations générales sur le rythme naturel du langage par *M. Stanislas Guyard*. Paris 1875 (Extrait du Journal as. 7 sér., t. 7. 8).

Die Rhetorik der Araber nach den wichtigsten Quellen dargestellt und mit angeführten Textauszügen nebst einem literaturgeschichtlichen Anhang versehen von Dr. *A. F. Mehren*. Kopenhagen 1853.

E. DICTIONARIES.

a. *Written by Orientals.*

*Ṣaḥāḥ al-'arabiyye (or aṣ-Ṣaḥāḥ) by *al-Ǧauharī* (Abū Naṣr Ismā'īl ibn Ḥammād † 393 H., beg. 10. Nov. 1002). 2 vols. Bulak 1282. 4⁰.

Lisān al-'arab by *al-Mukarram* (Ibn Manẓūr al-Ifrīkī al-Miṣrī al-Anṣārī al-Ḥazraǧī † 711 H., beg. 13. May 1311). 20 vols. 4⁰. Cairo 1308.

*al-Kāmūs al-muḥīṭ (or al-Kāmūs) by *al-Fīrūzābādī* († 816 or 817 H. = 1413/4). 2 vols. Calcutta 1817; 4 vols. Bulak 1279. 4⁰. id. 1301/2. — With Turkish Commentary 3 vols. Stambul 1272 and later. — *Commentary to the Kāmūs with the title Tāǧ-el-'arūs composed by *Sayyid Murtaḍā* az-Zubaidī († 1205 H., beg. 10 Sept. 1790). 10 vols. Cairo 1307.

Muḥīṭ al-muḥīṭ by *Buṭrus al-Bistānī*. 2 vols. Beirut 1286. (1869/70).

an-Nihāya fī ġarīb al-ḥadīṯ by *Ibn al-'Aṯīr* († 606 H., beg. 6. July 1209). 4 vols. Cairo 1311 (Dictionary to the Traditions).

Asās al-balāġa (Lexicographical Work, dealing esp. with the metaphorical meanings of words) by *az-Zamaḫšarī* († 538 H., beg. 16. July 1143). 2 vols. Bulak 1299.
Fiḳh al-luġa, Synonyms by *aṯ-Ṯa'ālibī* († 429 H., beg. 14. Oct. 1037). (Frequently reprinted; esp. in an expurgated edition Beirut 1888). Cf. Fleischer, Kleinere Schriften III, 152.
Ṯa'labs († 291 H. = 904) kitāb al-Faṣīḥ. Nach den Handschriften von Leiden, Berlin und Rom herausgegeben, mit kritischen und erläuternden Noten versehen von Dr. *J. Barth.* Leipzig 1876.
*Ǧawālīkī's al-Mu'arrab (a work on Arabic loan-words, by *al-Ǧawālīkī* † 465 H., beg. 17. Sept. 1072). Nach der Leydener Handschrift mit Erläuterungen herausgegeben von *Ed. Sachau.* Leipzig 1867. Cf. Z. d. D. Morg. Ges. 33, 208.
Liber as-Sojutii († 911 H., beg. 4. June 1505) de nominibus relativis, inscriptus Lubb al-lubab, arab. cum annot. crit. ed. *P. J. Veth.* 1—3. Lugduni Bat. 1840—51. 4⁰.
*Al-Moschtabih auctore Schamso'ddín Abu Abdallah Mohammed ibn Ahmed *ad-Dhababî* (aḏ-Ḏahabī † 748 H., beg. 13. April 1347). E codd. mss. editus a *P. de Jong.* Lugduni Batav. 1881. (On homonym proper names).
Kitābo-'l-adhdād sive liber de vocabulis arabicis quae plures habent significationes inter se oppositas auctore Abu Bekr *ibno-'l-Anbārī* († 328 H., beg. 18. Oct. 939) ed. *M. Th. Houtsma.* Lugduni Bat. 1881.

β *Written by Europeans.*

†*G. W. Freytag,* Lexicon Arabico-Latinum praesertim ex Djeuharii Firuzabadiique et aliorum libris confectum. Accedit index vocum latinorum locupletissimus. IV. Tomi. Hal. 1830—1837. 4⁰ maj.
G. W. Freytag, Lexicon Arabico-Latinum ex opere suo majore in usum tironum excerptum edidit. Halis 1836. 4⁰ maj.
*Maddu-l-Kamoos, an Arabic-English Lexicon derived from the best and the most copious eastern sources comprising a very large collection of words and significations omitted in the Kamoos, with supplements to its abridged and defective explanations, ample grammatical and critical comments, and examples in prose and verse: composed by means of the munificence of the most noble Algernon, Duke of Northumberland and the bounty of the British Government: by *Edward William Lane.* In two books: the first containing all the classical words and significations commonly known to the learned among the Arabs; the second, those that are of rare occurrence and not commonly known. Book I, Parts 1—5. London 1863—1874. Ed. by Stanley Lane Poole, Parts 6—8 (and Supplement) 1877—1893.

(From the letter ﻙ onwards, the book is incomplete; its continuation is not to be expected.)
*Supplément aux dictionnaires arabes par *R. Dozy*. 2 tom. Leyde 1881. — Cf. *Fleischer*, Studien über Dozy's Supplément: Berichte über die Verhandlungen der kgl. sächs. Ges. d. Wiss. zu Leipzig. Philol.-histor. Classe 1881—1887. Reprinted in Kleinere Schriften von H. L. Fleischer. Vol. II, pt. 1. Leipzig 1888. Vol. III id.
A. Kazimirski de Biberstein, Dictionnaire arabe-français I. II. Paris 1860.
†*A. Wahrmund*. Handwörterbuch der deutschen und neu-arabischen Sprache. I. Neuarabisch-deutscher Theil I, 1. 2. II, 1. 2. — II. Deutsch-neuarabischer Theil. Giessen 1870—77.
F. Steingass, The Student's Arabic-English Dictionary. London 1884.
H. Anthony Salmoné, An Arabic-English Dictionary on a new System. 2 vols. Vol. I Arabic-English; vol. II English Index. London 1890.
†Arabic-English Dictionary by the late *William Thomson Wortabet*. Second edition, revised and enlarged, Beyrout 1893.
George Percy Badger, English-Arabic Lexicon. London 1881.
F. Steingass, English-Arabic Dictionary for the use of both Travellers and Students. London 1882.
English-Arabic Dictionary by Mr. *J. Abcarius*. New edition revised and enlarged. Beyrout 1894.
†Vocabulaire arabe-français à l'usage des étudiants par un père missionnaire de la Cie de Jésus; 3. éd. Beyrouth 1893. (Arab.: al-Farāid ad-durrīye.)
Dictionnaire français-arabe par le P. *J.-B. Belot*, S. J. 2 parties. Beyrouth 1890.
*Die aramäischen Fremdwörter im Arabischen. Von *Siegmund Fränkel*. Leiden 1886.
Dictionnaire détaillé des noms des vêtements chez les Arabes. Par *R. Dozy*. Amsterdam 1845.
Die Namen der Säugethiere bei den südsemitischen Völkern. Von *Fritz Hommel*. Leipzig 1879.
Die Waffen der alten Araber aus ihren Dichtern dargestellt. Ein Beitrag zur arabischen Alterthumskunde, Synonymik und Lexicographie nebst Registern von *Friedrich Wilhelm Schwarzlose*. Leipzig 1886.
*Glossaire des mots espagnols et portugais dérivés de l'Arabe par *R. Dozy* et *W. H. Engelmann*. 2. éd. Leyde 1869.
Glossario etimologico de las palabras españolas de orígen oriental por *D. Leopoldo de Eguílaz y Yanguas*. Granada 1886.
Dictionnaire étymologique des mots français d'origine orientale par *Marcel Devic*. Paris 1876. — Cf. Remarques sur les mots français dérivés de l'Arabe par *Henri Lammens*. Beyrouth 1890.

F. KORAN, ISLAM, LIFE OF MUHAMMED. CHRISTIANITY.

a. Written by Orientals.

Al-Coranus seu Lex islamitica Muhammedis filii Abdallae Pseudoprophetae edita ex museo *Abrahami Hinckelmanni*. Hamburgi 1694.

Alcorani textus universus summa fide atque pulcherrimis characteribus descriptus, in latinum translatus, oppositis notis, auctore *Ludovico Marracio*. Patavii 1698 fol.

†Corani textus arabicus ad fidem librorum manuscriptorum et impressorum et ad praecipuorum interpretum lectiones et auctoritatem recensuit indicesque triginta sectionum et suratarum addidit *Gustavus Flügel*. Editio stereotypa C. Tauchnitzii. Tertium emendata; nova impressio Lipsiae 1869 (I. 1834; recensionis Flügelianae textum recognitum iterum exprimi curavit *Gustavus Mauritius Redslob*, Lipsiae 1837). (In Flügel's first edition and in numerous oriental editions of the Koran, the enumeration of the verses, which is indispensable for reference, is wanting).

*Concordantiae Corani arabicae. Ad literarum ordinem et verborum radices diligenter disposuit *Gustavus Flügel*. Editio stereotypa, Lipsiae 1842.

Chrestomathia Corani arabica, notas adjecit glossarium confecit *C. A. Nallino*. Lipsiae 1893.

al-Itḳān fī 'ulūm al-ḳur'ān, a sort of introduction to the Koran by *as-Suyūṭī* († 911 H., beg. 4. June 1505); 2 pts. Cairo 1278. — Sayúty's Itqán on the exegetic sciences of the Qorán. Edited by Mowlawies Basheerooddeen and Noorool-Haqq with an analysis by A. Sprenger. Calcutta 1852—54.

al-Kaššāf. Commentary on the Koran by *az-Zamaḥšarī* († 538 H., beg. 16. July 1143). 2 vols. Bulak 1281. — The Qoran with the commentary of Zamakhshari entitled the Kashshaf, an haqaiq al-tanzil, ed. by *W. Nassau Lees* and *Khadim Hosain* and *'Abd al-Hayi*. Calcutta 1856.

**Beidhawii* († 685 H., beg. 27. Febr. 1286; or 692) commentarius in Coranum ex codd.Parisiensibus Dresdensibus et Lipsiensibus edidit indicibusque instruxit *H. O. Fleischer*. 2 vol. Lipsiae 1846—48. 4⁰. — Indices ad Beidhawii commentarium in Coranum confecit *Winand Fell*. Leipzig 1878.

Chrestomathia Baidawiana. The commentary of El-Baidāwī on Sura III trans. and expld. . . . by *D. S. Margoliouth*. London 1895.

*Le Recueil des traditions musulmanes par Abou Abdallah ibn Ismail al-Bokhari (*al-Buḫārī* † 257 H., beg. 29. Nov. 870) publié par

L. Krehl. 1 – III. Leyde 1862—68 (incomplete). — Oriental edition: Ṣaḥīḥ al-Buḫārī. 8 vols. Cairo 1290; also frequently elsewhere, with and without commentary.

Ṣaḥīḥ *Muslim.* Collection of the Traditions of the Prophet, composed by Muslim († 261 H., beg. 16. Oct. 874). With commentary by *an-Nawawī* († 676 H., beg. 4. Juni 1277). 5 vols. Cairo 1283.

Maṣābīḥ as-sunna, composed by Ḥusain ibn Mas'ūd al-Farrā *al-Baġawī* († 516 H., beg. 12. March 1122). 2 vols. Cairo 1294.

Iḥyā al-'ulūm, by al-Gazālī († 505 H., beg. 10. Juli 1111). 4 vols. 4°. Bulak 1289. — (Cf. Richard Gosche, Über Ghazzâlis Leben und Werke: Abhdl. d. kgl. Akad. d. Wiss. zu Berlin 1858).

'*Abdu-r-razzāq's* Dictionary of the technical terms of the Sufies edited by *Aloys Sprenger.* Calcutta 1845.

*Das Leben Muhammeds nach Muhammed *ibn Isḥāk* († 151 H., beg. 26. Jan. 768) bearbeitet von 'Abd el-Malik *ibn Hischām* († 218 H., beg. 27. Jan. 833); hrsg. von *F. Wüstenfeld.* 2 Bände. Göttingen 1858—60. Oriental edition; Sīrat ibn Hišām. 2 vols. Cairo 1295. (Translated into German: Das Leben Muhammeds u. s. w. bearbeitet von *G. Weil.* Stuttgart 1864).

Muhammed in Medina. Das ist Vakidi's (al-*Wāḳidī* † 207 H., beg. 27. May 822) Kitab al-Maghazi in verkürzter deutscher Wiedergabe herausgegeben von *J. Wellhausen.* Berlin 1882.

Šamā'il *at-Tirmiḏī* († 279 H., beg. 3. April 892) Traditions respecting the Prophet. Cairo 1273; with commentary 2 vols. Bulak 1296.

Usd al-ġāba. List of 7500 persons who knew Muhammed, drawn up by *Ibn al-Aṯīr* († 630 H., beg. 18. Oct. 1232). 5 vols. Cairo 1286.

al-Iṣābe, A biographical dictionary of persons who knew Muhammed by *Ibn Hagar* (Ibn Ḥaǧar † 852 H., beg. 7. March 1448). Edited in Arabic by Mowlawies Mohammed Wajyh, 'Abdal-Ḥaqq, and Gholám Qádir and A. Sprenger. Bibliotheca Indica. Vol. I, Calcutta 1856; vol. IV, Calcutta 1873. Vol. II, fasc. 1—13; vol. III, fasc. 1—15.

Ḳiṣaṣ al-'anbiyā (Legends of the Prophet), by *aṯ-Ṯa'labī* († 427 H., beg. 5. Nov. 1035). Cairo 1297 and often.

Pillar of the creed of the Sunnites by *al-Nasafi,* ed. by *W. Cureton.* London 1843.

Ad-dourra al-fakhira: la perle précieuse de Ghazâlî (al-*Gazālī* † 505 H., beg. 10. July 1111) par *L. Gautier.* Genève 1878. — Muslim Eschatology.

Muhammedanische Eschatologie nach der Leipziger u. Dresdner Handschrift zum ersten Male arabisch und deutsch herausgegeben von *M. Wolff.* Leipzig 1872.

Disputatio pro religione Mohammedanorum adversus Christianos Textum arabicum (composed 942 H. = 1535) e codice Leidensi cum varr. lect. edidit *F. J. van den Ham.* Lugduni Bat. 1890.

Book of religious and philosophical sects by *Muhammed al-Shahrastáni (aš-Sahrastāni* † 528 H., beg. 29. March 1153). Now first edited by *W. Cureton*. 2 vol. London 1846. — Abu-'l-Fath' Muhammad asch-Schahrastàni's Religionsparteien und Philosophenschulen. Aus dem Arabischen übersetzt mit Anmerkungen von *Th. Haarbrücker*. 2 Bände. Halle 1850—1.
*(*Bible*) Kitāb al-mukaddas (Old Testament). London. R. Watts. 1822. (New Testament 1. vol. 1821.) — † Beirut, various editions. † New York 1867.
Arabic Bible-Chrestomathy with a Glossary edited by *Geo. Jacob*. Berlin 1888.

β *Written by Europeans.*

Der Koran nach Boysen von Neuem aus dem Arabischen übersetzt mit einer historischen Einleitung und Anmerkungen von *G. Wahl*. Halle 1828.
Der Koran. Aus dem Arabischen wortgetreu neu übersetzt mit Anmerkungen von *L. Ullmann*. 6. Aufl. 1862.
Le Koran, Traduction nouvelle, faite sur le texte arabe par Mr. *Kazimirski*. Nouv. éd. Paris 1854.
The Koran commonly called the Alcoran of Mohammed: translated into English from the Original Arabic. With explanatory notes taken from the most approved commentators. To which is prefixed a preliminary discourse. By *George Sale*. London 1774. Last ed. by E. M. Wherry "with additional notes and emendations". 4 vols. London 1882—87.
J. M. Rodwell, The Koran, translated from the Arabic. 2. ed. Lond. 1876.
The Qur'ân translated by *E. H. Palmer*. 2 parts. Oxford 1880. (The sacred books of the East translated by various oriental scholars and edited by F. Max Müller, vol. VI. IX).
Der Koran. Im Auszuge übersetzt von *Friedrich Rückert*, herausgegeben von *A. Müller*. Frankfurt a. M. 1888.
Die fünfzig ältesten Suren des Korans in gereimter deutscher Übersetzung von *M. Klamroth*. Hamburg 1800.
†*Geschichte des Qorâns von *Theodor Nöldeke*. Göttingen 1860.
Über die Religion der vorislamischen Araber. Eine zur Habilitation etc. öffentlich zu vertheidigende Abhandlung von *Ludolf Krehl*. Leipzig 1863.
*Skizzen und Vorarbeiten. Von *J. Wellhausen*. Drittes Heft. Reste arabischen Heidentumes. Berlin 1887.
Kinship and marriage in early Arabia. By *W. Robertson Smith*. Cambridge 1885.
*Das Leben und die Lehre des Mohammad nach bisher grösstentheils unbenutzten Quellen bearbeitet von *A. Sprenger*. Zweite Ausgabe. 3 Bände. Berlin 1869.

†Das Leben Muhammed's. Nach den Quellen populär dargestellt von *Theodor Nöldeke*. Hannover 1863.
*W. *Muir*, The Life of Mahomet and History of Islam. 4 vol. London 1858—61. 3rd edition 1 vol. 1894.
†Das Leben und die Lehre des Muhammed. Dargestellt von *Ludolf Krehl*. 1. Theil. Das Leben des Muhammed. Leipzig 1884.
Skizzen und Vorarbeiten von *J. Wellhausen*. Viertes Heft. 1. Medina vor dem Islam. 2. Muhammad's Gemeindeordnung von Medina. 3. Seine Schreiben, und die Gesandtschaften an ihn. Berlin 1889.
†Was hat Mohammed aus dem Judenthum aufgenommen? von *Abraham Geiger*. Bonn 1833.
*R. *Dozy*, Het Islamisme. Leiden 1863. 2 ed. Haarlem 1880; Essai sur l'histoire de l'Islamisme par R. Dozy trad. par V. Chauvin. Leyde-Paris 1879.
Snouck Hurgronje, Het mekkaansche Fest. Leiden 1880.
Die Mutaziliten oder die Freidenker im Islâm. Ein Beitrag zur allgemeinen Kulturgeschichte von *Heinrich Steiner*. Leipzig 1865.
De strijd over het Dogma in den Islâm tot op el-Ash'ari door Dr. *M. Th. Houtsma*. Leiden 1875.
Zur Geschichte Abu 'l-Hasan al-Aš'ari's († about 324 H. = 935) von *Wilhelm Spitta*. Leipzig 1876.
Exposé de la réforme de l'Islamisme commencée au IIIème siècle de l'Hégire par Abou-'l-Hasan Ali el-Ash'ari et continuée par son école. Avec des extraits du Texte arabe d'Ibn Asâkir par *M. A. F. Mehren*. Vol. II des Travaux de la 3e session du Congrès international des Orientalistes.
I. Goldziher, Die Schule der Zahiriten, ihr Ursprung, ihr System und ihre Geschichte. Leipzig 1884.
*Mohammedanische Studien von *I. Goldziher*. Erster Teil. Halle 1889. Zweiter Teil. Halle 1890.
Polemische und apologetische Literatur in arabischer Sprache zwischen Muslimen, Christen und Juden, nebst Anhängen verwandten Inhalts. Von *Moritz Steinschneider*. Abhandlungen für die Kunde des Morgenlandes VI, 3. Leipzig 1877.

G. JURISPRUDENCE.

al-Muwatta' fil-hadīt. Corpus juris composed by *Malik ibn Anas* al-Himyari al-Madanī († 179 H., beg. 27. March 795). Frequently printed; also with commentaries, e. g. that of az-Zarkāni († 1122 H., beg. 19. Febr. 1710). 4 vols. Bulak 1280.
Sunan Abī 'Abdallah al-Kazwīni, known as *Ibn Māǧa* († 273 H., beg. 8. June 886). Delhi 1282 and 1889. (Legal traditions).
Sunan *Abi Dā'ūd* Sulaimān as-Siǧistānī († 275 H., beg. 16. May 888); freq. printed, e. g. Bulak 1280. 2 vols. (Legal traditions).

al-Ǵāmiʿ by Abū ʿĪsā Muḥammad *at-Tirmiḏī* († 279 H., beg. 3. April 892). Frequently printed. (Legal traditions).
Sunan Abī 'Abd ar-raḥmān *an-Nasā'ī* († 303 H., beg. 17. July 915); lithogr. in Kanfūr 1847. (Legal traditions).
Flügel, Die Classen der hanefitischen Rechtsgelehrten: Abhandlungen der k. Sächs. Gesellschaft der Wissenschaften VIII. Leipzig 1860.
Jus Schafiiticum. At-Tanbīh auctore Abu Ishāk as-Shīrāzī (*Abū Ishāk aš-Šīrāzī* wrote the work in the year 452/3 H. = 1060/1) edidit A. W. T. Juynboll. Lugduni Bat. 1879.
Précis de Jurisprudence Musulmane selon le rite Châfeite, par Abu Chodjâ (*Abū Šuǵāʿ* † in the 6th cent. of the Flight). Publication du texte arabe, avec traduction et annotations, par S. Keijzer. Leyde 1859.
Minhādj aṭ-Ṭālibīn, le guide des zélés croyants. Manuel de jurisprudence musulmane selon le rite de Châfi'i (*aš-Šāfi'i*). Texte arabe, publié par ordre du gouvernement avec traduction et annotations par *L. W. C. van den Berg*. 3 vol. Batavia 1882—1884. (Cf. Snouck Hurgronje in the Indian Gids, 1884 ff. Elaborate criticism.)
Précis de jurisprudence musulmane suivant le rite malékite par *Sidi Khalil* (Ḥalīl lived in the 8th cent. of the Flight) publié par les soins de la Société asiatique. Quatrième édition. Paris 1877.
Maverdii (*al-Māwardī* † 450 H., beg. 28. Febr. 1058) constitutiones politicae. Ex recensione *Maximiliani Engeri*. Bonnae 1853.

H. PHILOSOPHY.

a Written by Orientals.

Documenta philosophiae Arabum, edidit latine vertit illustravit *Aug. Schmölders*. Bonnae 1836. — Cf. id. Essai sur les écoles philosophiques chez les Arabes et notamment sur la doctrine d'Algazzali. Paris 1842.
Tahāfut al-falāsifa (the mutual refutation of the philosophers) by *al-Ǵazāli* († 505 H., beg. 10. July 1111), *Ibn Rušd* († 595 H., beg. 3. Nov. 1198), *Hōǵa Zāde* († 893 H., beg. 17. Dec. 1487). Cairo 1303.
Die sogenannte Theologie des *Aristoteles* aus arabischen Handschriften zum ersten Male herausgegeben. Von *Fr. Dieterici*. Leipzig 1882 (Abhandlungen des Berl. Or.-Congresses). Cf. Die sogenannte Theologie des Aristoteles aus dem Arabischen übersetzt und mit Anmerkungen versehen von *Fr. Dieterici*. Leipzig 1883.
Il commento medio di *Averroë* alla Poetica di Aristotele pubbl. da *Fausto Lasinio*. Parte I. Il testo arabo: Annali della Università

Toscanè. Tomo XII. Pisa 1872. 4⁰. — Il testo arabo del commento medio di Averroe alla retorica di Aristotele, pubbl. da *Fausto Lasinio*. Firenze 1875. (Pubblicazioni del R. Istituto di studi superiori).

Alfārābī's († 950 A. D.) philosophische Abhandlungen aus Londoner, Leidener und Berliner Handschriften. Herausgegeben von *Friedrich Dieterici*. Leiden 1890. — Id. aus dem Arabischen übersetzt. Leiden 1892. — *Alfārābī's* Abhandlung der Musterstaat aus Londoner und Oxforder Handschriften herausgegeben von *F. Dieteerici*. Leiden 1895.

Philosophie und Theologie von Averroes (*Ibn Rušd* † 595 H., beg. 3. Nov. 1198). Herausgegeben von *M. J. Müller*. München 1859. — Aus dem Arabischen übersetzt. München 1875.

Le Guide des Égarés. Traité de Théologie et de Philosophie par Moïse ben Maïmoun dit *Maïmonide* († 605 H., beg. 16. July 1208). Publié pour la première fois dans l'original arabe et accompagné d'une traduction française par *Munk*. I—III. Paris 1856—66.

Kitāb Ihwān aṣ-ṣafā wa-hullān al-wafā (between 950—1000 of our era). 4 vols. Bombay 1305—1306. — A part of the rasāil Ihwān aṣ-ṣafā has also been printed in Cairo, 1306. — Die Abhandlungen der Ichwān Es-Safâ in Auswahl herausg. von *F. Dieterici*. 3 Hefte. Leipzig 1883—6.

Statio quinta et sexta et appendix libri Mevakif auctore ʿAdhad-eddin *el-Iǵî* († 756 H., beg. 16. Jan. 1355) cum commentario Ǵorǵānii ex codd. etc. edidit *Th. Sörensen*. Lipsiae 1848 (Scholastic Metaphysics).

Definitiones viri meritissimi Sejjid Scherif Ali ben Mohammed Dschordschani (*al-Ǵorǵānī* † 816 H., beg. 3. Apr. 1418). Accedunt definitiones theosophi Mohji-ed-din Mohammed ben Ali vulgo Ibn Arabi († 638 H., beg. 23 July 1240) dicti. Ed. et adnot. critica instruxit *Gustavus Flügel*. Lipsiae 1845.

β *Written by Europeans.*

Die griechischen Philosophen in der arabischen Überlieferung. Von *August Müller*. (Festschrift der Franckischen Stiftungen zu dem 50jährigen Doctorjubiläum Bernhardy's). Halle 1873.

Al-Kindî († ca. 850 A. D.) genannt „der Philosoph der Araber". Ein Vorbild seiner Zeit und seines Volkes. Von *G. Flügel*. Leipzig 1857. (Abhandlungen der D. Morg. Ges. 1. Band. Nr. 2). Cf. *Otto Loth, Al-Kindī* als Astrolog, Morgenländische Forschungen. Leipzig 1875, pp. 261 ff. and *Sir Wm. Muir, The Apology of Al-Kindy* 2 Ed. London 1887.

Al-Farabi, des arabischen Philosophen, Leben und Schriften. Von

Moritz Steinschneider: Mémoires de l'Académie Imp. des Sciences de St. Pétersbourg. VII. série, tome XIII, 4. 1869. 4⁰.
Ernest Renan, Averroès et l'Averroisme. 3. éd. Paris 1861.
Die Philosophie der Araber im X. Jahrhundert n. Chr. aus den Schriften der lauteren Brüder herausgegeben von *Fr. Dieterici*. Die Naturwissenschaft und Naturanschauung der Araber. Berlin 1861. — Die Propädeutik. Berlin 1865. — Die Logik und Psychologie. Leipzig 1868. — Die Anthropologie. Leipzig 1871. — Die Lehre von der Weltseele. Leipzig 1872. — Die Naturanschauung und Naturphilosophie. 2. Ausg. Leipzig 1876. — Einleitung und Makrokosmos. Leipzig 1876. — Mikrokosmos. Leipzig 1879.

I. NATURAL SCIENCE AND MEDICINE.

F. Wüstenfeld, Geschichte der arabischen Ärzte und Naturforscher. Göttingen 1840 (rather out of date).
Histoire de la médecine arabe par le Dr. *Lucien Leclerc*. 2 vol. Paris 1876 (insufficient).
Ibn Abī Useibia. Herausgegeben von *August Müller*. Königsberg i. Pr. 1884 (*Ibn Abī Uṣaibiʿa* † 668 H., beg. 14. May 1297 wrote this great work on the history of Arab physicians under the title: 'Uyūn al-ʾanbā' fī ṭabakāt al-ʾaṭibbā'. For which see Vol. II des travaux de la 6ᵉ session du Congrès international des Orientalistes à Leide. Leide 1884. p. 257 ff.).
Ḥayāt al-ḥaiwān (zoological work) by *ad-Damīrī* († 808 H., beg. 29. June 1405). 2 vols. Bulak 1284. Cairo 1305.
Kitāb al-kānūn fiṭ-ṭibb, Theory of Medicine, composed by Abu ʿAlī *ibn Sīnā* (*Avicenna* † 428 H., beg. 25. Oct. 1036). 3 vols. Bulak 1294.
al-Gāmiʿ li-mufradāt al-ʾadwiya wal-ʾagḏiya (On the common medicines and foods) by *Ḍiyāʾ ad-dīn Abū Muḥammad Ibn al-Baiṭār* († 646 H., beg. 26. April 1248). 4 vols. Bulak 1231.
Teḏkire (Science of medicine) by *Dāʾūd al-Antākī* († 1005 H., beg. 15. Aug. 1596). 3 vols. Cairo 1294.
La Chimie du moyen-âge . . . par *M. Berthelot*. Tome III. L'alchimie arabe comprenant une introduction et les traités de Cratès, d'el-Habib, d'Ostanès et de Djâber . . . texte et traduction . . avec la collaboration de *M. O. Houdas*. Paris 1893. 4⁰.
Matériaux pour servir à l'histoire des sciences mathématiques chez les Grecs et les Orientaux par *M. L. P. E. A. Sédillot*. 2 tomes. Paris 1845. 1849.
Traité des instruments astronomiques des Arabes, trad. par *J. J. Sédillot*. Paris 1834. 1835. Mémoires sur les instruments astronomiques des Arabes par *J. J. Sédillot*. Paris 1841—45.

K. HISTORY, BIOGRAPHY.

a *Written by Orientals*.

Ibn Coteiba's (*ibn Kutaiba* † 276 H. beg. 6. May 889) Handbuch der Geschichte herausgegeben von *Ferd. Wüstenfeld*. Göttingen 1850.— Oriental edition: Kitāb al-maʿārif. Cairo 1300.

Abu Bekr Muhammed ben al-Hasan *Ibn Doraid's* († 321 H., beg. 1. Jan. 933) genealogisch-etymologisches Handbuch herausgegeben von *F. Wüstenfeld*. Göttingen 1854.

*Chronologie orientalischer Völker von *Albêrûnî*. Herausgegeben von *Eduard Sachau*. Gedruckt auf Kosten der D. M. Ges. Leipzig 1878. 4⁰. — Chronology of ancient Nations. An English Version of the Arabic Text of the Athar ul Bâkiya of Albirûnî, or "Vestiges of the Past". Collected and reduced to writing by the Author in A. H. 390—1, A. D. 1000. Translated and Edited, with Notes and Index, by *C. E. Sachau*. Published for the Oriental Translation Fund of Great Britain and Ireland. Roy 8⁰. London 1879.

Ibn Wadhih (Wādih) qui dicitur *al-Jaʿqubī* (Yaʿḳūbī) Historiae (composed ca. 297 H.). 2 partes ed. *M. Th. Houtsma*. Lugduni Batav. 1883.

Anonyme Arabische Chronik Band XI vermuthlich das Buch der Verwandtschaft und Geschichte der Adligen von Abulhasan ahmed ben jahjā ben ğābir ben dāwūd elbelādori elbagdādi (*al-Balādurī* † 279 H., beg. 3. Apr. 893). Autogr. und herausgegeben von *W. Ahlwardt*. Greifswald 1883.

Kitāb al-aḫbār aṭ-ṭiwāl verf. von Abu Ḥanīfa Aḥmed ibn Dāūd ad-Dainawarī († 282 or 290 H.) hrs. von *Wladimir Girgas*. Leiden 1888.

*Annales auctore Abu Djafar Mohammed Ibn Djarir *At-Tabari* (aṭ-Ṭabarī † 309 H., beg. 12. May 921), quos ediderunt J. Barth, Th. Nöldeke, O. Loth (†), E. Prym, H. Thorbecke (†), S. Fränkel, D. H. Müller, M. Th. Houtsma, S. Guyard (†), V. Rosen et M. J. de Goeje I, 1—5; II, 1—3; III, 1—4. Leiden 1879 seq.

Maçoudi (*al-Masʿūdī* † 346 H., beg. 4. Apr. 957) Les prairies d'or. Texte et traduction par *C.Barbier de Meynard* et *Pavet de Courteille*. 9 tomes. Paris 1861—77. (id. 2 vols. Bulak 1283).

Hamzae Ispahanensis (*Hamza* wrote about 350 H.) annalium libri X. Edidit *J. M. E. Gottwaldt*. I. textus, II. transl. Petropoli-Lipsiae 1844. 1848.

Fragmenta historicorum arabicorum. Tomus primus continens partem tertiam operis Kitábo 'l-Oyun wa 'lhádáïk fi akhbári 'l-hadáïk (written after the 11th cent. A-D.) quem ediderunt M. J. de Goeje et P. de Jong. Lugduni Bat. 1868. 4⁰. — Tomus secundus continens partem operis Tadjáribo 'lOmami, auctore *Ibn Maskowaih* († 421 H., beg. 9. Jan. 1030) edidit M. J. de Goeje. Lugd. Bat. 1871.

*Ibn el-Athiri (*ibn al-'Atīr* † 630 H., beg. 18. Oct. 1232) Chronicon quod perfectissimum (el-Kāmil) inscribitur. Edidit *Carolus Johannes Tornberg*. 14 vol. Lugduni Bat. 1851—1876. — 12 vols. Bulak 1290 and later.

Commentaire historique sur le poème d'Ibn-Abdoun (*Ibn 'Abdūn* † 529 H., beg. 22. Oct. 1134) par Ibn Badroun (*Ibn Badrūn* wrote in the same century) publié par *R. P. A. Dozy*. Leide 1846 (Ouvrages arabes publiés par Dozy).

Historia saracenica arabice olim exarata a Georgio Elmacino (*al-Makīn* † 672 H., beg. 18. July 1273), edita et latine reddita opere et studiis *Thomae Erpenii*. Lugduni Bat. 1625.

Ta'rīh muhtasar ad-duwal (Outlines of History by Gregorius abū 'l-Farag Ibn el-'Ibri (*Barhebraeus* † 1286 A. D.) ed. by *Salhāni*. Beirut 1890. (The edition by *Pococke*, 2 tomi 4⁰. Oxonii 1663 is rare).

Elfachri. History of the Moslem Empires from the beginning to the end of the Califate by *Ibn etthiqthaqa* (wrote about 1302 A.D.). Edited in Arabic by *W. Ahlwardt*. Gotha 1860.

Abulfedae († 732 H., beg. 4. Oct. 1331). Annales muslemici arabice et latine. Opera et studiis *J. J. Reiskii*. nunc primum ed. *J. G. Ch. Adler*. 5 vol. Hafniae 1789—94. — 2 vols. Stambul 1286.

†Abulfedae historia Anteislamica, Arabice e duob. Codd. Paris. edidit, vers. lat. notis et indicibus auxit *H. O. Fleischer*. Lipsiae 1831. 4⁰.

Ta'rīh Zain ad-dīn *'Umar ibn al-Wardī* († 749 or 750 H. = 1348/9). 2 vols. Cairo 1285. — An excerpt: Aegyptus auctore Ibn al-Vardi. Edidit vertit notulisque illustravit *Martinus Frähn*. Halae 18ˑ4.

Ibn Haldūn († 808 H., beg. 29. June 1405) al-'ibar etc. History of the World. 7 vols. Bulak 1284. — Prolégomènes d'Ebn-Khaldoun. Texte arabe par *Quatremère*. 3 vols. Paris 1858 (Notices et extraits des mscr. XVI, 1. XVII, 1. XVIII, 1.). — Prolégomènes historiques d'Ibn Khaldoun. Traduction par *Mac Guckin de Slane*. 3 vols. Paris 1862—68 (Notices et extr. XIX, 1. XX, 1. XXI, 1).

The Tarikh al-Kholafá; or history of the Caliphs, from the death of Mohammad to the year 900 of the Hijrah by the celebrated Jalál al-Dín Al-Osyootí (*as-Suyūtī* † 911 H., beg. 4. June 1505). ed. by W. N. Lees und Mawlawi Abd al-Haqq. Calcutta 1857. Another edition Cairo 1305.

*Liber expugnationis regionum auctore Imámo Ahmed ibn Jahja ibn Djábir *al-Baladsori* (*al-Balādurī* † 279 H., beg. 3. Apr. 892) ed. *M. J. de Goeje*. Lugduni Bat. 1866. 4⁰.

Ousáma ibn Mounkidh un émir syrien au premier siècle des Croisades, (1095—1188) par *Hartwig Derenbourg*. Deuxième partie. Texte

arabe de l'autobiographie d'Ousâma. Paris 1886 (cf. Carlo de Landberg, Critica arabica II. Leyde 1888). — Ousâma ibn Mounkidh etc. par H. Derenbourg (French edition.). Paris 1889.
'Imād ed-dīn el-kātib el-isfahānī († 597 H. = 1201) Conquéte de la Syrie et de la Palestine par Salāh ed-dīn. Publié par le comte *Carlo de Landerg*. Vol. I. Texte arabe. Leyde 1888.
Vita et res gestae sultani Almalichi Alnasiri Saladini auctore Bohaddino F. Sjeddadi (*Bahā ad-dīn ibn Saddād* † 632 H. = 1234) edidit ac latine vertit *Albertus Schultens*. Lugduni Batav. 1732 (1755). fol.
Kitāb ar-raudatain fī ta'rīh ad-daulatain (History of Nureddin and Saladin) by Šihāb ad-dīn al-Mukaddasi, called *Abū Sāma* († 665 H. = 1267). Cairo. 2 vols. 1287.
Kitāb al-'ins al-galīl bi-ta'rīh al-kuds wal-halīl. History of Jerusalem and Hebron by *Mugīr ad-dīn* († 927 H., beg. 12. Dec. 1520). — Cf. Histoire de Jérusalem et d'Hébron. Fragments de la Chronique de Moudjir-ed-dyn traduits sur le texte arabe par *Henry Sauvaire*. Paris 1876.
Die Chroniken der Stadt Mekka. Gesammelt und herausgegeben von *Ferdinand Wüstenfeld* (I Azraki. II Fākihi, Fāsī, Ibn Dhuheira. III Kutb ed-dīn. IV German edition). I—IV. Leipzig 1857—61.
Hulāsat al-wafā bi'ahbār dār al-mustafā (History of the town of Medīna) by *as-Samhūdi* († 911 H., beg. 4. June 1505). Bulak 1285. — Extracts translated by *Wüstenfeld* in den Abhandlungen der k. Ges. der Wissenschaften zu Göttingen. Bd. IX. 1860.
*al-Hitat (Geography and History of Egypt) by *al-Makrīzī* († 845 H., beg. 22. May 1441). 2 vols. Bulak 1270. — Histoire des Sultans Mamlouks de l'Egypte, écrite en arabe par Taki-eddin-Ahmed Makrizi, traduite en français et accompagnée de notes par *Quatremère*. 2 vol. Paris 1837—45. 4⁰.
Abūl-Mahāsin ibn Tagri Bardii († 874 H., beg. 11. July 1469) Annales (History of Egypt) I, 1. 2 ediderunt *T. G. J. Juynboll* et *B. F. Matthes*. II, 1. 2. ed. *T. G. J. Juynboll*. Lugduni Bat. 1852—61 (incomplete).
Husn al-muhādara. History of Egypt by *as-Suyūtī* († 911 H., beg. 4. June 1505). 2 vols. Cairo.
'Agāib al-ātār fit-tarāgim wal-ahbār (History of Egypt) by *al-Gabartī* († 1236 = 1821). 4 vols. Cairo *n. d.*
Ahmedis Arabsiadae (*Ahmed ibn 'Arabšah* † 854 H., beg. 14. Febr. 1450) vitae et rerum gestarum Timuri, qui vulgo Tamerlanes dicitur historia. (Ed.) Latine vertit etc. *S. H. Manger*. 2 vol. Leovardiae 1767. 1772. — Cairo 1285.
The History of the Almohades by Abdo-'l-Wāhid *al-Marrekoshī* (wrote in the year 621 H. = 1224) edited by *R. Dozy*. 2. ed. Leyden 1881.

Historia Abbadidarum praemissis scriptorum Arabum de ea dynastia locis nunc primum editis, auctore *R. P. A. Dozy*. I—III. Lugduni Bat. 1849. 4⁰. (Deals w. Spain).

Annales regum Mauretaniae a condito Idrisidarum imperio ad annum fugae 726, ab Abu-l Hasan Ali ben Abd Allah Ibn Abi Zer' Fesano, vel ut alii malunt Abu Muhammed Salih ibn Abd el Halim Granatensi conscriptos ed. illustr. *Carolus Joh. Tornberg.* 2 vol. Upsaliae 1843. 1846.

Histoire de l'Afrique et de l'Espagne intitulée al-Bayáno 'l-Moghrib par *Ibn Adhári* (de Maroc) (*Ibn al-'Idārī* wrote between 363 and 366 H.) et fragments de la chronique d'Arib (de Cordoue) publiés par *R. P. A. Dozy*. 2. vols. Leyde 1848—51.

Analectes sur l'histoire et la littérature des Arabes d'Espagne par *Al-Makkari* (*al-Makkarī* † 1041 H., beg. 30. July 1631). Publiés par *R. Dozy, G. Dugat, L. Krehl* et *W. Wright*. 2 vol. Leyde 1855—61. (Conf. *Fleischer*, Textverbesserungen in Al-Makkari's Geschichtswerke. Kleinere Schriften. Vol. II pt. 1. Leipzig 1888.) — Lettre à M. Fleischer contenant les remarques critiques et explicatives sur le texte d'Al-Makkari par *R. Dozy*. Leyde 1871. — Cf. The history of the Mohammedan Dynasties in Spain by Ahmed ibn Mohammed Al-Makkarī. Translated and illustrated by Pascual de Gayangos. 2 vol. London 1840—3. 4⁰.

Bibliotheca arabo-sicula, ossia Raccolta di testi arabici che toccano la geografia, la storia, la biografia e la bibliografia della Sicilia, messi insieme da *Michele Amari*. Lipsia 1857; Appendice, ibid. 1875.

Alberuni's India, an account of the religion, philosophy, literature, chronology, astronomy, customs, laws and astrology of India about 1030. Ed. by *Edw. Sachau.* London 1887. 4⁰. — Id. An English edition with notes and indices. By *E. Sachau.* London. 2 vol. 1888.

Scriptorum Arabum de Rebus Indicis loci et opuscula inedita rec. et illustr. *Joannes Gildemeister.* Fasc. pr. Bonnae 1838. — Cf. id., Dissertationis de rebus Indiae, quo modo in Arabum notitiam venerint, pars I. Bonnae 1838.

Ibn Challican, Vitae illustrium virorum. E codd. nunc primum arabice edidit variis lectionibus, indicibusque locupletissimis instruxit *Ferd. Wüstenfeld.* Gottingae 1835—40, 4⁰. — Ibn Hallikān († 681 H., beg. 11. Apr. 1282). 2 vols. Bulak 1275; another edition 1299. — Ibn Khallikan's Biographical Dictionary, translated from the Arabic by Baron *Mac Guckin de Slane.* 4 vol. Paris-London 1843—71. 4⁰.

Fawāt' al-wafayāt (supplement to Ibn Hallikān) by *aṣ-Ṣalāḥ ul-Kutubī* († 764 H., beg. 21. Oct. 1362). 2 vols. Bulak 1283.

The biographical dictionary of illustrious men chiefly at the beginning of Islamism by Abu Zakariya Jahya el-*Nawawi* († 676 H. = 1277) edited by *Ferd. Wüstenfeld.* Göttingen 1842—47 (cf. *idem* for the Life and Writings of el-Nawawi, Göttingen 1849, from the 4th vol. of the Abhandl. d. kgl. Ges. d. Wiss. zu Gött.).
Nuzhat al-'alubba fi ṭabakāt al-'udabā. Concerning celebrated Men, By Abul-Barakāt *al-'Anbāri* († 577 H., beg. 17. May 1181). Cairo lithogr. n. d.

β *Written by Europeans.*

†Vergleichungstabellen der muhammedanischen und christlichen Zeitrechnung nach den ersten Tagen jedes muhammedanischen Monats berechnet. Herausgegeben von *Ferd. Wüstenfeld.* Leipzig 1844.— Fortsetzung der Wüstenf. Vergl.-Tab. bis 1500 von *E. Mahler.* Leipzig 1887.

*Die Geschichtsschreiber der Araber und ihre Werke. Von *F. Wüstenfeld.* (From the XXVIII. and XXIX. vol. of the Abhandlungen der Kgl. Ges. d. W. zu Göttingen). Göttingen 1882. 4⁰.

*Genealogische Tabellen der Arabischen Stämme und Familien . . . Aus den Quellen zusammengestellt von *Ferdinand Wüstenfeld.* Göttingen 1852. q.-fol. — Register zu den genealogischen Tabellen der Arabischen Stämme und Familien. Mit historischen und geographischen Bemerkungen von *Ferdinand Wüstenfeld.* Göttingen 1853.

Caussin de Perceval, Essai sur l'histoire des Arabes avant l'islamisme 3 vol. Paris 1847.

Geschichte der Perser und Araber zur Zeit der Sassaniden. Aus der arabischen Chronik des Tabari übersetzt und mit ausführlichen Erläuterungen und Ergänzungen versehen von *Th. Nöldeke.* Leyden 1879.

†*Der Islam im Morgen-f und Abendland. Von *A. Müller.* 2 Bände. Berlin 1885. 1887. (Allgemeine Geschichte in Einzeldarstellungen hrsgg. von L. Oncken. Zweite Hauptabteilung. Vierter Teil).

*Geschichte der Chalifen. Nach handschriftlichen grösstenteils noch unbenützten Quellen bearbeitet von *Gustav Weil.* 3 Bände. Mannheim 1846—51. — Geschichte des Abbasidenchalifats in Aegypten. Von *Gustav Weil.* 2 Bände. Stuttgart 1860—2.

†Geschichte der islamitischen Völker von Mohammed bis zur Zeit des Sultan Selim übersichtlich dargestellt von *Gustav Weil.* Stuttgart 1866.

†Geschichte der Araber bis auf den Sturz des Chalifats von Bagdad. Von *Gustav Flügel.* 2. Aufl. Leipzig 1864.

The Caliphate, its rise, decline, and fall from original sources by *Sir William Muir.* London 1891. New and revised edition 1894.

Handbuch der morgenländischen Münzkunde. Von *J. G. Stickel.* 2 Hefte. Leipzig 1865—70. 4⁰.
Catalogue of Oriental Coins in the British Museum, 9 vol. London 1875—1889.
The Mohammadan Dynasties, chronological and genealogical Tables with historical Introductions by *St. Lane-Poole.* London 1894.
Die Charidschiten unter den ersten Omayyaden. Ein Beitrag zur Geschichte des ersten islamischen Jahrhunderts von *R. E. Brünnow.* Leiden 1884.
De opkomst der Abbasiden in Chorasan door *G. van Vlooten.* Leiden 1890.
Mémoires sur les Carmathes du Bahrain et les Fatimides par *M. J. de Goeje.* Leiden 1886.
Die Statthalter von Ägypten zur Zeit der Chalifen. Von *F. Wüstenfeld.* Parts 1 and 2. Abhandlungen der Kgl. Ges. d. Wissenschaften zu Göttingen. 1875 (4⁰). Band 20. Parts 3 and 4. ibid. 1876, Band 21.
History of the Moors in Spain to the Conquest of Andalusia by the Almoravides (711—1110), by *R. Dozy.* German Edition with additions by the Author. 2 vols. Leipzig 1874.
Poesie und Kunst der Araber in Spanien und Sicilien. Von *Adolf Friedrich von Schack.* 2 Bände. Berlin 1865. 2. Aufl. 1877.
*Culturgeschichte des Orients unter den Chalifen. Von *Alfred von Kremer.* 2 Bände. Wien 1875—77.
Das Einnahmebudget des Abbasiden-Reichs vom Jahre 360 H. (918—919) von *Alfred von Kremer.* Denkschriften der philos.-hist. Classe der Kais. Akademie der Wiss. in Wien. Bd. XXXVI. 1887.
*Geschichte der herrschenden Ideen des Islams. Der Gottesbegriff, die Prophetie und Staatsidee. Von *Alfred v. Kremer.* Leipzig 1868.

Die Baustile. Historische und technische Entwicklung. Des Handbuchs der Architectur (von *J. Durm*) Zweiter Theil. 3. Band, zweite Hälfte; Die Baukunst des Islam. Von *Franz Pascha.* Darmstadt 1887.
Prisse d'Avennes, L'art arabe d'après les monuments du Caire depuis le VIIe siècle jusqu' à la fin du XVIIIe. 3 vol. fol. 1 vol. 4. Paris 1877. — La décoration arabe. (Extrait du grand ouvrage.) Paris 1865. fol.

L. COSMOGRAPHY, GEOGRAPHY, ETHNOGRAPHY, TRAVELS.

a Written by Orientals.

Cosmographie de Chems ed-din Abou Abdallah Mahommed *ed-Dimichqi* (*ad-Dimiški* † 654 H., beg. 30. Jan. 1256). Texte arabe publié d'après l'édition commencée par M. Frähn, et d'après les manuscrits par *M. A. F. Mehren*. St. Pétersbourg 1866. 4⁰. — Manuel de la cosmographie du moyen âge, traduit de l'arabe „Nokhbet ed-dahr fi 'adjaib-il-birr wal-bah'r" de Shems ed-dîn Abou-'Abdallah Mohammed de Damas et accompagnée d'éclaircissements par *M. A. F. Mehren*. Copenhague 1874.

*Zakarija Ben Muhammed ben Mahmúd el-*Cazwini's* (*al-Kazwīnī* † 682 H., beg. 1. Apr. 1283) Kosmographie. Herausg. von *Ferd. Wüstenfeld*. 2 Bände. Göttingen 1848—9. — id. nach der Wüstenfeld'schen Textausgabe etc. übersetzt von *Hermann Ethé*. Erster Halbband. Leipzig 1868.

Harīdat al-'ağāib wa-farīdat al-ġarāib, a species of Cosmography composed by '*Umar ibn al-Wardī* († 749 or 750 H. = 1348 or 9). Cairo 1292.

Specimen e literis orientalibus exhibens *az-Zamaksarīi*, (*az-Zamaḫšarī* † 538 H., beg. 16. July 1143) lexicon geographicum quod auspice T. G. J. Juynboll edidit *Mathias Salverda de Grave*. Lugduni Bat. 1856.

Al-Hamdānī's († 334 H., beg. 13. Aug. 945) Geographie der Arabischen Halbinsel. Nach den Handschr. herausgegeben von *David Heinrich Müller*. Leiden 1884.

Das geographische Wörterbuch des Abu 'Obeid 'Abdallah ben 'Abd el-'Azīz *el-Bekri* († 487 H. = 1094) nach den Handschriften zu Leiden, Cambridge, London und Mailand herausgegeben von *Ferd. Wüstenfeld*. 2 Bände. Göttingen, Paris 1876. 1877.

*Jacut's (*Yāḳūt* † 626 H. = 1229) Geographisches Wörterbuch aus den Handschriften zu Berlin, St. Petersburg und Paris auf Kosten der Deutschen Morgenländischen Gesellschaft herausgegeben von *Ferdinand Wüstenfeld*. 6 Bände. Leipzig 1866—73.

Jacut's Moschtarik, das ist: Lexicon geographischer Homonyme. Herausgegeben von *Ferd. Wüstenfeld*. Göttingen 1846.

Marāṣid al-iṭṭilā'i, Lexicon geographicum ed *T. G. J. Juynboll* I—VI. Lugduni B. 1850—64. (An Extract from Yāḳūt).

Géographie d'Aboulféda (*Abū'l-fidā* † 732 H., beg. 4. Oct. 1331). Texte arabe par *Reinaud* et *Mac-Guckin de Slane*. Paris 1840. — Géographie d'Ismaïl Abou 'l-Fédā en arabe publiée par *Charles Schier*. Éd. autogr. Dresde 1846. — Géographie d'Aboulféda, traduite de l'arabe en français par *Reinaud* I (*Introduction générale à la géographie des Orientaux) II, 1 Paris 1848; II, 2 par *Stanislas Guyard*. Paris 1883.

*Bibliotheca geographorum arabicorum. Edidit *M. J. de Goeje*.

Pars prima. Viae regnorum. Descriptio ditionis moslemicae auctore Abu-Ishāk al-Fārisī *al-Istakhri* (*al-Iṣṭaḫrī*, cf. Zeitschrft d. D. Morgenl. Ges. Bd. 25, p. 42 ff.). Lugduni Bat. 1870.

Pars secunda. Viae et regna. Descriptio ditionis moslemicae auctore Abu 'l-Kāsim *Ibn Ḥaukal* (ibid.). Lugduni Bat. 1873.

Pars tertia. Descriptio imperii Moslemici auctore *Al-Mokaddasi (al-Mukaddasī* wrote in year 378 the H.). Lugduni Bat. 1876. Cf. Description of Syria &c. by Mukaddasi. Translated from the Arabic by *Guy Le Strange*. (Palestine Pilgrims' Text Society). Pars quarta. Continens indices, glossarium et addenda et emendanda ad part. I—III auctore *M. J. de Goeje*. Lugduni Bat. 1879.

Pars quinta. Compendium libri Kitāb al-boldān auctore *Ibn al-Fakīh* al-Hamadhani (wrote ca. A. D. 290). Lugd. Bat. 1885.

Pars sexta. Kitâb al-masālik wal-mamālik (liber viarum et regnorum) auctore Abu'l-Kâsim Obaidallah ibn Abdallah *ibn Khordādbeh* (Ibn Hordādbeh wrote in the second half of the 9th cent. A. D.) et excerpta e Kitâb al-Kharâdj (K. al-harağ Taxbook) auctore *Kodāma ibn Dja'far* (*Kudāma* ibn Ğa'far wrote about 930 A. D.). Lugduni Bat. 1889.

Pars septima. Kitāb al-a'lāk an-nafisa VII auctore Abū Alī Ahmed ibn Omar *ibn Rosteh* (wrote before 301. H.) et Kitâb al-boldān auctore Ahmed ibn abī Ja'kūb ibn Wādhih al-Kātib *al-Ja'kūbī* (cf. p. 157). Lugduni Bat. 1892.

Pars octava. Kitâb at-tanbih wa'l-ischrâf auctore *al-Masûdî* (cf. p. 157). Accedunt indices et glossarium ad tomos VII et VIII. Lugduni Bat. 1894.

Description de l'Afrique et de l'Espagne par *Edrīsī* (wrote 548 H., beg. 29 March 1153) texte arabe publié pour la première fois d'après les man. de Paris et d'Oxford avec une traduction, des notes et un glossaire par *R. Dozy* et *M. J. de Goeje*. Leyde 1866.

The travels of *Ibn Jubair* (*Ibn Gubair* end of the 6th cent.) edited by *William Wright*. Leyden 1852.

Voyages d'Ibn Batoutah (*Ibn Batūṭa* † 779 H., beg. 10 May 1377). Texte arabe, accompagnée d'une traduction par *C. Defrémery* et *B. R. Sanguinetti* (Publications de la Société asiatique). 4 vol. Paris 1853—58; deux. tir. 1874—77. — Cairo 1288.

β *Written by Europeans.*

F. *Wüstenfeld*, Die Litteratur der Erdbeschreibung bei den Arabern. Zeitschrift für vergleichende Erdkunde hrsgg. von J. G. Lüdde I, 1841, S. 24—67.

Carte générale des provinces européennes et asiatiques de l'Empire Ottoman, dressée par *Henri Kiepert* 4 feuilles. Deux. éd. entièrement corrigée et augmentée d'un index alphabetique. Berlin 1892.

(Karte von) Arabien zu C. Ritters Erdkunde, Buch III, West-Asien, Teil XII und XIII bearbeitet von *H. Kiepert*. Neue berichtigte Ausgabe, die Orthographie revidiert von *Th. Nöldeke*. Berlin 1867 (D. Reimer).

Skizze der Geschichte und Geographie Arabiens von den ältesten
Zeiten bis zum Propheten Muḥammad. Auf Gruud der Inschriften,
der Angaben der alten Autoren und der Bibel von *Eduard Glaser.*
Zweiter Band. Berlin 1890.

Die alte Geographie Arabiens als Grundlage der Entwicklungsgeschichte
des Semitismus von *A. Sprenger.* Bern 1875.

Arabien im sechsten Jahrhundert. Eine ethnographische Skizze von
Otto Blau. Mit einer Karte: Zeitschrift der deutschen morgenl.
Gesellschaft. Leipzig 1869 (XXIII B.) p. 559—592.

Arabien und die Araber seit hundert Jahren. Eine geographische
und geschichtliche Skizze von *Albrecht Zehme.* Halle 1875.

Palestine under the Moslems. A description of Syria and the Holy
Land from A. D. 650 to 1500. Translated from the works of the
mediæval Arab Geographers by *Guy le Strange.* (London) 1890.

Relation de l'Égypte par *Abdallatif* ('Abd al-Laṭîf al-Baġdâdi † 629 H.,
beg. 29. Oct. 1231). Le tout traduit et enrichi de notes par
Silvester de Sacy. Paris 1810. 4⁰. (The text of 'Abd al-Laṭîf
has been published by *J. White:* 'Abdollatiphi Historiae Aegypti
compendium. Oxonii 1800).

*Beschreibung von Arabien. Aus eigenen Beobachtungen und im
Lande selbst gesammelten Nachrichten abgefasst von *Carsten
Niebuhr.* Kopenhagen 1772. 4⁰.

Carsten Niebuhrs Reisebeschreibung nach Arabien und andern um-
liegenden Ländern. 1. Band. Kopenhagen 1774. 2. Band. 1778;
English edtn. 2 vols. Edinb. 1792.

†*Travels in Arabia (1814) comprehending an account of those territories
in Hedjaz which the Mohammedans regard as sacred. By the
late *John Lewis Burckhardt.* London, 2 vol. 1829. — *Johann
Ludwig Burckhardt's* Reisen in Arabien, enthaltend eine Beschreibung derjenigen Gebiete in Hedjaz, welche die Mohammedaner
für heilig achten ... Aus dem Englischen übersetzt. Weimar 1830.

†*J. L. Burckhardt,* Notes on the Bedouins and Wahábys. 2 vol.
London 1831. — Bemerkungen über die Beduinen und Wahabi's.
Weimar 1831.

Richard Burton, Personal narrative of a pilgrimage to El Medinah and
Meccah. 2 vol. London 1857 (and frequently, also in the Tauchnitz
edition).

*Travels in Arabia Deserta by *Charles M. Doughty.* 2 vol. Cambridge
1888. (With new map).

Adolf von Wrede's Reise in Ḥadhramaut, Beled Beny 'Issâ und Beled
el Ḥadschar. Herausgegeben ... von *H. Freiherr von Maltzan.*
Braunschweig 1870. — Reise nach Südarabien und Geographische
Forschungen im und über den südwestlichen Teil Arabiens von
Heinrich Freihern von Maltzan. Braunschweig 1873.

Mekka von Dr. *C. Snouck Hurgronje*. 2 Bände. Mit Bilder-Atlas. Haag 1888. 1889.
†*An account of the manners and customs of the modern Egyptians, written in Egypt etc. By *Edward William Lane*. Various editions. London. — Lane, Sitten und Gebräuche der heutigen Egypter. Übersetzt von *J. Zenker*. 3 Bde. Leipzig 1852.
E. W. Lane, Arabian society in the middle ages. Studies from the Thousand and One Nights ed. by *Stanley Lane Poole*. London 1883. (Supplement to the "Manners and Customs", containing the notes to Lane's translation of the Thousand and One Nights (v. *infra*).

M. VERSE.

Delectus veterum carminum arabicorum. Carmina selegit et edidit *Th. Noeldeke*, glossarium confecit *A. Müller*. Berolini 1890.
Über Poesie und Poetik der Araber von *Wilhelm Ahlwardt*. Gotha 1856. 4⁰.
Beiträge zur Kenntniss der Poesie der alten Araber. Von *Th. Nöldeke*. Hannover 1864.
Kitāb al-agānī by Abu 'l-Farag 'Alī al-*Isfahānī* († 352 H., beg. 30. Jan. 962). 20 vols. Bulak 1285. — Alii Ispahanensis liber cantilenarum magnus, ed. *Kosegarten*. T. 1. Gripesvoldiae 1840. 4⁰. — The twenty-first volume of The Kitâb al-aghâni ed. by *Rud. E. Brünnow*, Leyden 1888. — Tables alphabetiques du Kitâb al-Agânî par *J. Guidi*. 1er fasc. Leide 1895.
Kitāb raudat al-adab fī tabakāt šu'arā' al-'arab by *Iskander-Aga Abkarius* (modern Beyrout scholar). Beirut 1858.
Ḥizānat al-adab wa-lubb lubāb lisān al-'arab, by '*Ab-dal-Kādir ibn 'Umar al-Baǵdādī* († 1093 H.; beg. 21. Aug. 1629) 4 vols. Bulak 1291 (A work on poets; on the margin are printed the Šawāhid al-'Ainī). An index to the poets appeared from the pen of *Guidi* in the transactions of the R. Accademia dei Lincei, Rome 1887.
*The Diwans of the six ancient Arabic poets Ennābiga, 'Antara, Tharafa, Zuhair, 'Alqama and Imruulqais, ed. by *W. Ahlwardt*. London. 1870.
Bemerkungen über die Ächtheit der alten Arabischen Gedichte mit besonderer Beziehung auf die sechs Dichter etc. von *W. Ahlwardt*. Greifswald 1872.
Le Dîwân de *Nâbiga* Dhobyânî publié par *H. Dérenbourg*. Journal asiatique 1868—9.
H. Thorbecke, 'Antarah, ein vorislamischer Dichter. Leipzig 1867.
Die Gedichte des '*Alkama* Alfaḥl. Mit Anmerkungen herausgegeben von *Albert Socin*. Leipzig 1867.
Le diwan d'*Amro'lkais* par le Bon *de Slane*. Paris 1837. 4⁰. With

Commentary by al-Baṭalyūsi. Cairo 1308. Cf. Amrilkais, der Dichter und König. Von *Fr. Rückert*. Stuttgart und Tübingen 1843.

†*Septem *Moʿallakât* carmina antiquissima Arabum, textum etc. rec. *F. A. Arnold*. Lipsiae 1850 (out of print) — With commentary by *az-Zauzani* († 375 H., beg. 24. May 958). Cairo 1288. A commentary by Abû Zakariyā Sahya *at-Tibrizi* († 420 H., beg. 11. Aug. 1108) on ten ancient Arabic poems edited from the Mss. of Cambridge, London and Leiden by Charles James Lyall. Fasc. I Bibliotheca Indica, New Series, No. 789, Calcutta 1891; Fasc. II ib. No. 840. Calc. 1894.

Der Diwan des *Lebīd*. Nach einer Handschrift zum ersten Male herausgegeben von *Jūsuf Dijā-ad-dīn al-Châlidī*. Wien 1880. Cf. *A. von Kremer* in den Sitzungsberichten der phil.-hist. Classe der Kais. Akademie d. Wissenschaften 98. Bd. 2 Heft. Wien 1881. — Die Gedichte des Lebīd. Nach der Wiener Ausgabe übersetzt und mit Anmerkungen versehen aus dem Nachlasse des *Dr. A. Huber* herausgegeben von *Carl Brockelmann*. Leiden 1891.

Die Mufaḍḍalījāt (Anthology of the Grammarian *al-Mufaḍḍal*; † about 170 H.) Nach den Handschriften herausgegeben von *Heinrich Thorbecke*. Erstes Heft. Leipzig 1885.

*Hamasae carmina cum Tebrisii scholiis integris edidit, indicibus instruxit, versione latina et commentario illustr. *G. G. Freytag*. 2 vol. Bonnae 1828—47 (collected by Abu Tammām † 190, beg. 27. Nov. 805; at-Tabrīzī Comm. † 420 H., beg. 11. Aug. 1108). Another edition Bulak 1296. Cf. Hamâsa oder die ältesten arabischen Volkslieder, gesammelt von Abu Temmâm, übersetzt und erläutert von *Friedrich Rückert*. 2 T. Stuttgart 1846.

The Hudsailian poems contained in the manuscript of Leyden edited in Arabic and translated with annotations by *J. G. L. Kosegarten*. Vol. I. London 1854. 4⁰. — Letzter Theil der Lieder der Hudhailiten, arabisch und deutsch: Skizzen und Vorarbeiten von *J. Wellhausen*. 1. Heft. Berlin 1884. Comp. Z. der D. Morgenl. Gesellschaft 39. pp. 104, 151, 411 ff.

Die Gedichte des *'Urwa ibn Alward*. Von *Th. Nöldeke*: Abhandlungen der Kgl. Ges. d. Wiss. zu Göttingen. Hist.-Phil. Classe 11. Gedichte und Fragmente des *'Aus ibn Hajar*, gesammelt, herausgegeben und übersetzt von *Rudolf Geyer*: Sitzungsberichte der Kais. Akademie der Wissenschaften in Wien. Philos.-hist. Classe. Band 126. Wien 1892.

Anīs al-ǵulasā' fī dīwān *al-Hansā'* (The poetess al-Ḥansā is said to have died A. H. 24, beg. 7. Nov. 644 A. D.) Beirut 1888. — Le diwan d'al Hansa' traduit par le *P. de Coppier* et suivi de fragments inédits d'Al-Ḥirniq. Beyrouth 1889.

Ibn Hišāmi († 762 H., beg. 11 Nov. 1360) Commentarius in Carmen *Ka'bi ben Zoheir* Bānat Su'ād ed. *Guidi.* Lipsiae 1871. 1874.

Der Dīwān des Garwal b. Aus *al-Huṭej'a* († between 68—70 H.) Bearbeitet von *Ignaz Goldziher*: Zeitschrift der D. Morgenl. Gesellschaft Bd. 46, S. 1—53; 173—225; 471—527; Bd. 47, S. 43—85; 163—201. Also in a collected edition. Leipzig 1893.

Dīwān sayyidnā *Ḥassān ibn Ṯābit* († 54 H., beg. 30. Aug. 683). Tunis 1281.

Dīwân d'*al-Ahṭal*, Texte arabe publié pour la première fois d'après le manuscrit de St. Pétersbourg et annoté par le P. A. Salhani S. J. Beyrouth 1891.

Divan de *Férazdak* († 110 H., beg. 16. April 728) récits de Mohammed ben-Habib d'après Ibn-el-Arabi publié sur le manuscrit de Sainte-Sophie de Constantinople avec une traduction française par *R. Boucher*. Paris 1870. 4⁰. (incomplete).

Maǧmū' muštamil 'ala hams dawāwīn (an-Nābiga, 'Urwa, Hātim, 'Alkama and Farazdak) Cairo 1293 cf. Z. der D. Morgenl. Gesellschaft 31, 667 ff.

Chalef elahmar's (died after 155 H.) Qasside. Berichtigter arabischer Text etc. von *A. Ahlwardt*. Greifswald 1859.

Dīwān *al-Buḥturī* († 190 H., beg. 27 Nov. 805). Constantinople 1300.

Diwan des Abu Nowas nach der Wiener und Berliner Handschrift mit Benutzung anderer Handschriften herausgegeben von *W. Ahlwardt*. 1. Die Weinlieder. Greifswald 1861. — Dīwān Abī Nuwās. Cairo 1277. (*Abū Nuwās* † about 195 H. = 810).

Diwan poëtae Abu-'l-Walīd Moslim ibno-'l-Walīd al-Anṣāri cognomine *Ṣario-'l-ghawāni* (*Ṣarī al-ǧawāni* † 208 H., beg. 16. May 823) quem edidit *M. de J. Goeje*. Lugduni Bat. 1875. 4⁰.

Al-anwār az-zāhiya fī dīwān Abi'l-'Atāhiya (*Abu'l-'Atāhiya* † 221 H., beg. 26. Dec. 835). Beirut 1886. 2me édit. 1888.

Dīwān *Abī Tammām* Habīb ibn Aus aṭ-Ṭā'ī († 231 H., beg. 7. Sept. 845). Cairo 1292.

Dīwān amīr al-mu'minīn *Ibn-al-Mu'tazz* al-'Abbāsi († 296 H. = 909) Cairo 1891. Cf. Über Leben und Werk des 'Abdallah ibn al-Mu'tazz von *Otto Loth*. Leipzig 1882.

Mutanabbii (*al-Mutanabbī* † 354 H. = 965) carmina cum commentario Waḥidii primum edidit, indicibus instruxit, varias lectiones adnotavit *Fr. Dieterici*. Berolini 1861. 4⁰.

Dīwān *Abī Firās* al-Ḥamdāni († 357 beg. 7. Dec. 967). Beirut 1873.

Abu'l-'Alā' al-Mā'arrī († 449 H., beg. 10 March 1057) Sakṭ ez-zind, Poems with Commentary. 2 vols. Bulāk 1286 and 1302 (Another edition Beirut 1884). — Luzūm mā lā yalzam. Bombay 1303. 4⁰; Luzūmiyāt 2 vols. Cairo 1891. — *Caroli Rieu* de Abul-Alae

poetae arabici vita et carminibus. Bonnae 1843. Cf. Zeitschrift der D. Morgenl. Gesellschaft 29, 304; 30, 40; 31, 471 ff.
Yatīmat ad-dahr fī šuʻarā' ahl al-ʻaṣr, Anthology composed by Abū Manṣūr ʻAbd al-Malik at-Ṯaʻālibī († 429 H., beg. 14. Oct. 1037) 4 vols. Damascus 1302.
Anthologie arabe ou choix de poésies arabes inédites traduites pour la première fois en français et accompagnées d'observations critiques et littéraires par *M. Grangeret de La Grange*. (Paris) 1828.

N. BELLES-LETTRES, ETHICS, ROMANCES.

*The Kāmil of *El-Mubarrad* († 285 H., beg. 28. Jan. 898), edited for the German Oriental Society by *W. Wright*. Part. 1—12; Leipzig 1864—92. A reprint appeared in Cairo 2 vols. 1308.
al-ʻIḳd al-farīd, by *Ibn ʻAbd-rabbihi al-Andalusī* († 328 H., beg. 28. March 860) 3 vols. Bulak 1293.
Kitāb al-Muwaššā of Abu 't-Ṭayyib Muḥammed ibn Isḥāq *al-Waššā* (lived 860—936 A. D.) edited by *R. Brünnow*. Leyden 1886.
Ibn *Arabschah* († 854 H., beg. 14. Febr. 1450) Fructus imperatorum et jocatio ingeniosorum edidit *G. G. Freytag*. 2. vol. Bonnae 1832. 4⁰. — Oriental editions with the title: Fākihat al-ḫulafā' wa-mufākahat aẓ-ẓurafā'.
Makāmāt badīʻ az-zamān *al-Hamadānī* (al-Hamadāni, the predecessor of Ḥariri died 398 H., beg. 17. Sept. 1007) with commentary by Šeiḫ Muḥammad Abdo. Beirut 1889. Other Makamat of Hamadāni Constantinople 1298.
*Les séances de *Ḥariri* (al-Ḥarīrī † 516 H., beg. 12. March 1122), avec un commentaire choisi pao *Silvestre de Sacy*; 1 éd. Paris 1822; 2. éd. par *Reinaud* et *J. Derenbourg*. 2 tom. Paris 1847—1853. — With the Commentary of *aš-Šarīšī* († 619 H., beg. 15. Feb. 1222) 2 vols. Bulak 1284. — Makāmāt (Vowelled text) 2. Ed. Beyrouth 1886. — The Assemblies of Al-Hariri, transl. &c. by Thomas Chenery. Vol 1 1867. — *Do*. Arabic text with English notes &c. by F. Steingass 1895.
*Kitāb *Adab al-Kātib* (proply. an aid to elegant writing) composed by Muḥammed Abdallāh ibn Muslim *Ibn Kutaiba* († in the 2ⁿᵈ. half of the 3ʳᵈ. Centy. of the Flight). Cairo 1300.
Kitāb al-matal as-sā'ir fī 'ādāb al-kātib waš-šāʻir (Treatise on Style) by *Ibn al-Aṯīr al-Ǧazarī* († 637 H., beg. 3. Aug. 1239) Cairo 1282.
Rasāil (Letters) abi'l-Faḍl badīʻ az-zamān al-*Hamadāni* († 398 H., beg. 17. Sept. 1007). Constantinople 1298.
al-Maidānī († 518 H., beg. 19. Feb. 1124) Maǧmaʻ al-amtāl. (Collection of Proverbs). 2 vols. Bulak 1284. — Arabum proverbia,

vocalibus instruxit, latine vertit, commentario illustravit *G. G. Freytag* I, II, III (a b.), Bonnae 1838—43.

†Les colliers d'or, allocutions morales de Zamakhschari (*az-Zamaḫšarī* † 538 H., beg. 16. July 1143) texte arabe suivi d'une traduction française et d'un commentaire philologique par *C. Barbier de Meynard*. Paris 1876.

Ali's hundert Sprüche arabisch und persisch paraphrasiert von Reschideddin Watwat, nebst einem doppelten Anhang arabischer Sprüche herausgegeben, übersetzt und mit Anmerkungen begleitet von *H. L. Fleischer*, Leipzig 1837. 4⁰.

Sirāǧ al-mulūk (Ethics and Anecdotes) composed by Abū Bekr Muḥammed *aṭ-Ṭarṭūšī* al-Mālikī († 520 H., beg. 27. Jan. 1126). Cairo 1289.

Muḥāḍarāt al-ᵓudabā wa-muḥāwarāt aš-šuʿarāʾ wal-bulaġāʾ, a species of Ethics with Anecdotes by *ar-Rāġib al-Iṣfahānī* († in the beginning of the 6th centy. of the Flight). 2 vols. Cairo 1287. 4⁰.

al-Mustaṭraf fī kull fann al-mustaẓraf, a species of anthological Encyclopaedia compiled by *Aḥmad al-Ilšihī* (lived about 800 H.) 2 vols. Cairo 1304. 1307.

Sīret ʿAntar ibn Saddād, 32 vols. Cairo 1286. 1307. (another recension 10 vols. Beirut 1871). Cf. Antar, a Bedoueen romance. Translated from Arabic by *T. Hamilton*. Part. I, i—iv. London 1820.

Alf laila wa-laila. Tausend und eine Nacht arabisch. Nach einer Handschrift aus Tunis herausg. von *Maximilian Habicht* I—VIII; fortges. von *H. L. Fleischer* IX—XII vol. Breslau 1825—43. (This edition is not suitable for beginners in Arabic, as the language is in many parts strongly influenced by the vulgar tongue). — The Alif Laila or book of the thousand nights and one night, published from an Egyptian Ms. by *W. H. Macnaghten*. 4 vols. Calcutta 1839—42. — 4 vols. Bulak 1279. — Original in expurgated edition. Beyrout 1888—90. — Following the earlier Bulak edition: The thousand and one nights commonly called, in England, The Arabian nights' entertainments. Translated by *W. Lane*. 3 vol. London. 1 ed. 1841. Other editions by *Edw. Stanley Poole* (the last 1882).

PART II.

PARADIGMS, CHRESTOMATHY
AND
GLOSSARY.

Socin, Arabic Grammar.³   A

PARADIGMATA.

TABULA I.
Suffixa et Praefixa in flexione verbi adhibita.

Persona	Numerus	Perfectum	Imperfectum
3. masc.	sing.	ـَ	(يَ) يَـ ـُ
3. fem.	”	ـَتْ	(تَ) تَـ ـُ
2. masc.	”	ـْتَ	(تَ) تَـ ـُ
2. fem.	”	ـْتِ	(تَ) تَـ ـِينَ (ـِى)
1.	”	ـْتُ	(أَ) أَـ ـُ
3. masc.	dual.	ـَا	(يَ) يَـ ـَانِ (ـَا)
3. fem.	”	ـَتَا	(تَ) تَـ ـَانِ (ـَا)
2.	”	ـْتُمَا	(تَ) تَـ ـَانِ (ـَا)
3. masc.	plur.	ـُوا	(يَ) يَـ ـُونَ (ـُوا)
3. fem.	”	ـْنَ	(يَ) يَـ ـْنَ
2. masc.	”	ـْتُمْ	(تَ) تَـ ـُونَ (ـُوا)
2. fem.	”	ـْتُنَّ	(تَ) تَـ ـْنَ
1.	”	ـْنَا	(نَ) نَـ ـُ

TABULA II.
Paradigma flexionis verbi sani stirpis I.
Activum

Persona	Numerus	Perfectum	Imperfectum					Imperativus
			Indicativus	Subjunctivus	Apocopat.	Energ. I.	Energ. II	
3. masc.	sing.	قَتَلَ	يَقْتُلُ	يَقْتُلَ	يَقْتُلْ	يَقْتُلَنْ	يَقْتُلَنَّ	
3. fem.	”	قَتَلَتْ	تَقْتُلُ	تَقْتُلَ	تَقْتُلْ	تَقْتُلَنْ	تَقْتُلَنَّ	
2. masc.	”	قَتَلْتَ	تَقْتُلُ	تَقْتُلَ	تَقْتُلْ	تَقْتُلَنْ	تَقْتُلَنَّ	اُقْتُلْ
2. fem.	”	قَتَلْتِ	تَقْتُلِينَ	تَقْتُلِي	تَقْتُلِي	تَقْتُلِنْ	تَقْتُلِنَّ	اُقْتُلِي
1.	”	قَتَلْتُ	اَقْتُلُ	اَقْتُلَ	اَقْتُلْ	اَقْتُلَنْ	اَقْتُلَنَّ	

					اِقْتُلْ	اِقْتُلُوا اِقْتُلْنَ
			يَقْتُلْ	يَقْتُلَ	يَقْتُلُ	قَتَلَ
3. masc. dual		يَقْتُلَا	يَقْتُلَا	يَقْتُلَا	يَقْتُلَانِ	قَتَلَا
3. fem. "		تَقْتُلَا	تَقْتُلَا	تَقْتُلَا	تَقْتُلَانِ	قَتَلَتَا
2. "		تَقْتُلَا	تَقْتُلَا	تَقْتُلَا	تَقْتُلَانِ	قَتَلْتُمَا
3. masc. plur.		يَقْتُلُوا	يَقْتُلُوا	يَقْتُلُوا	يَقْتُلُونَ	قَتَلُوا
3. fem. "		يَقْتُلْنَ	يَقْتُلْنَ	يَقْتُلْنَ	يَقْتُلْنَ	قَتَلْنَ
2. masc. "		تَقْتُلُوا	تَقْتُلُوا	تَقْتُلُوا	تَقْتُلُونَ	قَتَلْتُمْ
2. fem. "		تَقْتُلْنَ	تَقْتُلْنَ	تَقْتُلْنَ	تَقْتُلْنَ	قَتَلْتُنَّ
1. "		نَقْتُلْ	نَقْتُلَ	نَقْتُلُ	قَتَلْنَا

TABULA III.
Paradigma flexionis
Passivi I verbi sani

Persona	Numerus	Perfectum	Imperfectum		
			Indicativus	Subjunctivus	Apocopat.
3. masc.	sing.	قُتِلَ	يُقْتَلُ	يُقْتَلَ	يُقْتَلْ
3. fem.	”	قُتِلَتْ	تُقْتَلُ	تُقْتَلَ	تُقْتَلْ
2. masc.	”	قُتِلْتَ	تُقْتَلُ	تُقْتَلَ	تُقْتَلْ
2. fem.	”	قُتِلْتِ	تُقْتَلِينَ	تُقْتَلِي	تُقْتَلِي
1.	”	قُتِلْتُ	أُقْتَلُ	أُقْتَلَ	أُقْتَلْ
3. masc.	dual.	قُتِلَا	يُقْتَلَانِ	يُقْتَلَا	يُقْتَلَا
3. fem.	”	قُتِلَتَا	تُقْتَلَانِ	تُقْتَلَا	تُقْتَلَا
2.	”	قُتِلْتُمَا	تُقْتَلَانِ	تُقْتَلَا	تُقْتَلَا
3. masc.	plur.	قُتِلُوا	يُقْتَلُونَ	يُقْتَلُوا	يُقْتَلُوا
3. fem.	”	قُتِلْنَ	يُقْتَلْنَ	يُقْتَلْنَ	يُقْتَلْنَ
2. masc.	”	قُتِلْتُمْ	تُقْتَلُونَ	تُقْتَلُوا	تُقْتَلُوا
2. fem.	”	قُتِلْتُنَّ	تُقْتَلْنَ	تُقْتَلْنَ	تُقْتَلْنَ
1.	”	قُتِلْنَا	نُقْتَلُ	نُقْتَلَ	نُقْتَلْ

PARADIGMATA.

TABULA IV.
Paradigma stirpium verbi quadrilitteralis.

Stirps	Genus	Perfectum	Imperfectum	Imperativus	Participium	Infinitivus
I	Act.	ܩܰܛܶܢ	ܢܩܰܛܶܢ	ܩܰܛܶܢ	ܡܩܰܛܶܢ	ܩܰܛܳܢܽܘ
I	Pass.	ܩܰܛܰܢ	ܢܩܰܛܰܢ		ܡܩܰܛܰܢ	
II	Act.	ܐܶܬܩܰܛܰܢ	ܢܶܬܩܰܛܰܢ	ܐܶܬܩܰܛܰܢ	ܡܶܬܩܰܛܰܢ	ܡܶܬܩܰܛܳܢܽܘ
II	Pass.	ܐܶܬܩܰܛܰܢ	ܢܶܬܩܰܛܰܢ		ܡܶܬܩܰܛܰܢ	

TABULA V.
Paradigma stirpium verbi sani.

	I	II	III	IV	V
Perfectum Activi	قَتَلَ	قَتَّلَ	قَاتَلَ	أَقْتَلَ	تَقَتَّلَ
Imperfectum "	يَقْتُلُ	يُقَتِّلُ	يُقَاتِلُ	يُقْتِلُ	يَتَقَتَّلُ
Imperativus "	اُقْتُلْ	قَتِّلْ	قَاتِلْ	أَقْتِلْ	تَقَتَّلْ
Participium "	قَاتِلٌ	مُقَتِّلٌ	مُقَاتِلٌ	مُقْتِلٌ	مُتَقَتِّلٌ
Perfectum Passivi	قُتِلَ	قُتِّلَ	قُوتِلَ	أُقْتِلَ	تُقُتِّلَ
Imperfectum "	يُقْتَلُ	يُقَتَّلُ	يُقَاتَلُ	يُقْتَلُ	يُتَقَتَّلُ
Participium "	مَقْتُولٌ	مُقَتَّلٌ	مُقَاتَلٌ	مُقْتَلٌ	مُتَقَتَّلٌ
Infinitivus	قَتْلٌ	تَقْتِيلٌ	قِتَالٌ vel مُقَاتَلَةٌ	إِقْتَالٌ	تَقَتُّلٌ

PARADIGMATA.

		VI	VII	VIII	IX	X
Perfectum	Activi	تَفَاعَلَ	اِفْتَعَلَ	اِنْفَعَلَ	اِفْعَلَّ	اِسْتَفْعَلَ
Imperfectum	"	يَتَفَاعَلُ	يَفْتَعِلُ	يَنْفَعِلُ	يَفْعَلُّ	يَسْتَفْعِلُ
Imperativus	"	تَفَاعَلْ	اِفْتَعِلْ	اِنْفَعِلْ	اِفْعَلِلْ	اِسْتَفْعِلْ
Participium	"	مُتَفَاعِلٌ	مُفْتَعِلٌ	مُنْفَعِلٌ	مُفْعَلٌّ	مُسْتَفْعِلٌ
Perfectum	Passivi	تُفُوعِلَ	اُفْتُعِلَ	اُنْفُعِلَ		اُسْتُفْعِلَ
Imperfectum	"	يُتَفَاعَلُ	يُفْتَعَلُ	يُنْفَعَلُ		يُسْتَفْعَلُ
Participium	"	مُتَفَاعَلٌ	مُفْتَعَلٌ	مُنْفَعَلٌ		مُسْتَفْعَلٌ
Infinitivus	"	تَفَاعُلٌ	اِفْتِعَالٌ	اِنْفِعَالٌ	اِفْعِلَالٌ	اِسْتِفْعَالٌ

TABULA VI.
Paradigma flexionis
Activi I verbi mediae geminatae

Persona	Numerus	Perfectum	Imperfectum			Imperativus
			Indicativus	Subjunctivus	Apocopatus	
3. masc.	sing.	فَرَّ	يَفِرُّ	يَفِرَّ	يَفْرِرْ يَفِرَّ	
3. fem.	„	فَرَّتْ	تَفِرُّ	تَفِرَّ	تَفْرِرْ تَفِرَّ	
2. masc.	„	فَرَرْتَ	تَفِرُّ	تَفِرَّ	تَفْرِرْ تَفِرَّ	اِفْرِرْ فِرَّ
2. fem.	„	فَرَرْتِ	تَفِرِّينَ	تَفِرِّي	تَفِرِّي	فِرِّي
1.	„	فَرَرْتُ	أَفِرُّ	أَفِرَّ	أَفْرِرْ أَفِرَّ	
3. masc.	dual.	فَرَّا	يَفِرَّانِ	يَفِرَّا	يَفِرَّا	
3. fem.	„	فَرَّتَا	تَفِرَّانِ	تَفِرَّا	تَفِرَّا	
2.	„	فَرَرْتُمَا	تَفِرَّانِ	تَفِرَّا	تَفِرَّا	فِرَّا
3. masc.	plur.	فَرُّوا	يَفِرُّونَ	يَفِرُّوا	يَفِرُّوا	
3. fem.	„	فَرَرْنَ	يَفْرِرْنَ	يَفْرِرْنَ	يَفْرِرْنَ	
2. masc.	„	فَرَرْتُمْ	تَفِرُّونَ	تَفِرُّوا	تَفِرُّوا	فِرُّوا
2. fem.	„	فَرَرْتُنَّ	تَفْرِرْنَ	تَفْرِرْنَ	تَفْرِرْنَ	اِفْرِرْنَ
1.	„	فَرَرْنَا	نَفِرُّ	نَفِرَّ	نَفْرِرْ نَفِرَّ	

TABULA VII.
Paradigma flexionis
Passivi I verbi mediae geminatae

Persona	Numerus	Perfectum	Imperfectum		
			Indicativus	Subjunctivus	Apocopatus
3. masc.	sing.	فُرَّ	يُفَرُّ	يُفَرَّ	يُفْرَرْ يُفَرَّ
3. fem.	„	فُرَّتْ	تُفَرُّ	تُفَرَّ	تُفْرَرْ تُفَرَّ
2. masc.	„	فُرِرْتَ	تُفَرُّ	تُفَرَّ	تُفْرَرْ تُفَرَّ
2. fem.	„	فُرِرْتِ	تُفَرِّينَ	تُفَرِّي	تُفَرِّي
1.	„	فُرِرْتُ	أُفَرُّ	أُفَرَّ	أُفْرَرْ أُفَرَّ
3. masc.	dual.	فُرَّا	يُفَرَّانِ	يُفَرَّا	يُفَرَّا
3. fem.	„	فُرَّتَا	تُفَرَّانِ	تُفَرَّا	تُفَرَّا
2.	„	فُرِرْتُمَا	تُفَرَّانِ	تُفَرَّا	تُفَرَّا
3. masc.	plur.	فُرُّوا	يُفَرُّونَ	يُفَرُّوا	يُفَرُّوا
3. fem.	„	فُرِرْنَ	يُفْرَرْنَ	يُفْرَرْنَ	يُفْرَرْنَ
2. masc.	„	فُرِرْتُمْ	تُفَرُّونَ	تُفَرُّوا	تُفَرُّوا
2. fem.	„	فُرِرْتُنَّ	تُفْرَرْنَ	تُفْرَرْنَ	تُفْرَرْنَ
1.	„	فُرِرْنَا	نُفَرُّ	نُفَرَّ	نُفْرَرْ نُفَرَّ

TABULA VIII
Paradigma stirpium verbi mediae geminatae contractarum.

	I	III	IV	VI	VII	VIII	X
Perfectum Activi	فَرَّ	فَارَّ	فَرَّرَ	تَفَارَّ	اِنْفَرَّ	اِفْتَرَّ	اِسْتَفَرَّ
Imperfectum "	يَفِرُّ	يُفَارُّ	يُفَرِّرُ	يَتَفَارُّ	يَنْفَرُّ	يَفْتَرُّ	يَسْتَفِرُّ
Imperativus "	اُفْرِرْ	فَارَّ	فَرِّرْ	تَفَارَّ	اِنْفَرَّ	اِفْتَرَّ	اِسْتَفْرِرْ
Participium "	فَارٌّ	مُفَارٌّ	مُفَرِّرٌ	مُتَفَارٌّ	مُنْفَرٌّ	مُفْتَرٌّ	مُسْتَفِرٌّ
Perfectum Passivi	فُرَّ	فُورَّ	فُرِّرَ	تُفُورَّ	اُنْفُرَّ	اُفْتُرَّ	اُسْتُفِرَّ
Imperfectum "	يُفَرُّ	يُفَارُّ	يُفَرَّرُ	يُتَفَارُّ	يُنْفَرُّ	يُفْتَرُّ	يُسْتَفَرُّ
Participium "	مَفْرُورٌ	مُفَارٌّ	مُفَرَّرٌ	مُتَفَارٌّ	مُنْفَرٌّ	مُفْتَرٌّ	مُسْتَفَرٌّ
Infinitivus "	فَرٌّ	فِرَارٌ	تَفْرِيرٌ	تَفَارٌّ	اِنْفِرَارٌ	اِفْتِرَارٌ	اِسْتِفْرَارٌ

TABULA IX.
Paradigma formarum selectarum flexionis
verborum hamzatorum

	Verbi pr. ء	Verbi sec. ء	verbi tert. ء
I. Perf. Act.	أَثَرَ	كَثِبَ لَأَمَ	قَرَأَ
Impf. „	يَأْثِرُ	يَكْثُبُ يَلْأَمُ	يَقْرَأُ
Imperat. „	اِيثِرْ (اُءْءُدْ)	اِكْثُبْ اِلْأَمْ	اِقْرَأْ
Partic. „	آثِرٌ	لَآئِمٌ	قَارِئٌ
Perf. Pass.	أُثِرَ	لُئِمَ	قُرِئَ
Imperf. „	يُؤْثَرُ	يُلْأَمُ	يُقْرَأُ
II. Imperf. Act.	يُؤَثِّرُ	يُلَئِّمُ	يُقَرِّئُ
Infin. „	تَأْثِيرٌ	تَلْئِيمٌ	تَفْرِئَةٌ
IV. Perf. Act.	آثَرَ	أَلْأَمَ	أَقْرَأَ
Perf. Pass.	أُوثِرَ	أُلْئِمَ	أُقْرِئَ
VIII. Perf. Act.	اِيتَثَرَ (اِتَّخَذَ)	اِلْتَأَمَ	اِقْتَرَأَ
Imperf. „	يَأْتَثِرُ (يَتَّخِذُ)	يَلْتَئِمُ	يَقْتَرِئُ
Perf. Pass.	أُوتِثرَ (اُتُّخِذَ)	اُلْتُئِمَ	اُقْتُرِئَ
Imperf. „	يُوتَثَرُ (يُتَّخَذُ)	يُلْتَأَمُ	يُقْتَرَأُ

TABULA X.
Paradigma flexionis verborum
primae radicalis و et ي

		Verbi pr. و Imperf. i	Verbi pr. و Imperf. a	Verbi pr. و sani	Verbi pr. ي
I.	Perf. Act.	وَصَلَ	وَدَعَ	وَسِخَ	يَسَرَ
	Imperf. „	يَصِلُ	يَدَعُ	يَوْسَخُ	يَيْسِرُ
	Imperat. „	صِلْ	دَعْ	(اَوْسَنْ ,اِيجَلْ)	اِيسِرْ
	Imperf. Pass.	يُوصَلُ	يُودَعُ	يُوسَخُ	يُوسَرُ
	Infinit.	صِلَةٌ	دَعَةٌ, وَدْعٌ	وَسَخٌ	يَسْرٌ
IV.	Perf. Act.	أَوْصَلَ	أَوْدَعَ	أَوْسَخَ	أَيْسَرَ
	Imperf. „	يُوصِلُ	يُودِعُ	يُوسِخُ	يُوسِرُ
	Partic. „	مُوصِلٌ	مُودِعٌ	مُوسِخٌ	مُوسِرٌ
	Infinit.	إِيصَالٌ	إِيدَاعٌ	إِيسَاخٌ	إِيسَارٌ
VIII.	Perf. Act.	اِتَّصَلَ	اِتَّدَعَ	اِتَّسَخَ	اِتَّسَرَ
	Imperf. „	يَتَّصِلُ	يَتَّدِعُ	يَتَّسِخُ	يَتَّسِرُ
	Perf. Pass.	اُتُّصِلَ	اُتُّدِعَ	اُتُّسِخَ	اُتُّسِرَ
X.	Perf. Act.	اِسْتَوْصَلَ	اِسْتَوْدَعَ	اِسْتَوْسَخَ	اِسْتَيْسَرَ
	Infinit.	اِسْتِيصَالٌ	اِسْتِيدَاعٌ	اِسْتِيسَاخٌ	اِسْتِيسَارٌ

TABULA XI.
Paradigma flexionis
Activi I verbi mediae radicalis و

Persona	Numerus	Perfectum	Imperfectum			Imperativus
			Indicativus	Subjunctivus	Apocopatus	
3. masc.	sing.	قَالَ	يَقُولُ	يَقُولَ	يَقُلْ	
3. fem.	"	قَالَتْ	تَقُولُ	تَقُولَ	تَقُلْ	
2. masc.	"	قُلْتَ	تَقُولُ	تَقُولَ	تَقُلْ	قُلْ
2. fem.	"	قُلْتِ	تَقُولِينَ	تَقُولِي	تَقُولِي	قُولِي
1.	"	قُلْتُ	أَقُولُ	أَقُولَ	أَقُلْ	
3. masc.	dual.	قَالَا	يَقُولَانِ	يَقُولَا	يَقُولَا	
3. fem.	"	قَالَتَا	تَقُولَانِ	تَقُولَا	تَقُولَا	
2.	"	قُلْتُمَا	تَقُولَانِ	تَقُولَا	تَقُولَا	قُولَا
3. masc.	plur.	قَالُوا	يَقُولُونَ	يَقُولُوا	يَقُولُوا	
3. fem.	"	قُلْنَ	يَقُلْنَ	يَقُلْنَ	يَقُلْنَ	
2. masc.	"	قُلْتُمْ	تَقُولُونَ	تَقُولُوا	تَقُولُوا	قُولُوا
2. fem.	"	قُلْتُنَّ	تَقُلْنَ	تَقُلْنَ	تَقُلْنَ	قُلْنَ
1.	"	قُلْنَا	نَقُولُ	نَقُولَ	نَقُلْ	

TABULA XII.
Paradigma flexionis
Activi I verbi mediae radicalis ى

Persona	Numerus	Perfectum	Imperfectum			Imperativus
			Indicativus	Subjunctivus	Apocopatus	
3. masc.	sing.	سَارَ	يَسِيرُ	يَسِيرَ	يَسِرْ	
3. fem.	”	سَارَتْ	تَسِيرُ	تَسِيرَ	تَسِرْ	
2. masc.	”	سِرْتَ	تَسِيرُ	تَسِيرَ	تَسِرْ	سِرْ
2. fem.	”	سِرْتِ	تَسِيرِينَ	تَسِيرِى	تَسِيرِى	سِيرِى
1.	”	سِرْتُ	أَسِيرُ	أَسِيرَ	أَسِرْ	
3. masc.	dual.	سَارَا	يَسِيرَانِ	يَسِيرَا	يَسِيرَا	
3. fem.	”	سَارَتَا	تَسِيرَانِ	تَسِيرَا	تَسِيرَا	
2.	”	سِرْتُمَا	تَسِيرَانِ	تَسِيرَا	تَسِيرَا	سِيرَا
3. masc.	plur.	سَارُوا	يَسِيرُونَ	يَسِيرُوا	يَسِيرُوا	
3. fem.	”	سِرْنَ	يَسِرْنَ	يَسِرْنَ	يَسِرْنَ	
2. masc.	”	سِرْتُمْ	تَسِيرُونَ	تَسِيرُوا	تَسِيرُوا	سِيرُوا
2. fem.	”	سِرْتُنَّ	تَسِرْنَ	تَسِرْنَ	تَسِرْنَ	سِرْنَ
1.	”	سِرْنَا	نَسِيرُ	نَسِيرَ	نَسِرْ	

TABULA XIII.
Paradigma flexionis
Passivi I verbi mediae radicalis , vel ى

Persona	Numerus	Perfectum	Imperfectum		
			Indicativus	Subjunctivus	Apocopatus
3. masc.	sing.	قِيلَ	يُقَالُ	يُقَالَ	يُقَلْ
3. fem.	”	قِيلَتْ	تُقَالُ	تُقَالَ	تُقَلْ
2. masc.	”	قِلْتَ	تُقَالُ	تُقَالَ	تُقَلْ
2. fem.	”	قِلْتِ	تُقَالِينَ	تُقَالِي	تُقَالِي
1.	”	قِلْتُ	أُقَالُ	أُقَالَ	أُقَلْ
3. masc.	dual.	قِيلَا	يُقَالَانِ	يُقَالَا	يُقَالَا
3. fem.	”	قِيلَتَا	تُقَالَانِ	تُقَالَا	تُقَالَا
2.	”	قِلْتُمَا	تُقَالَانِ	تُقَالَا	تُقَالَا
3. masc.	plur.	قِيلُوا	يُقَالُونَ	يُقَالُوا	يُقَالُوا
3. fem.	”	قِلْنَ	يُقَلْنَ	يُقَلْنَ	يُقَلْنَ
2. masc.	”	قِلْتُمْ	تُقَالُونَ	تُقَالُوا	تُقَالُوا
2. fem.	”	قِلْتُنَّ	تُقَلْنَ	تُقَلْنَ	تُقَلْنَ
1.	”	قِلْنَا	نُقَالُ	نُقَالَ	نُقَلْ

Socin, Arabic Grammar.²

TABULA XIV.
Paradigma stirpium verborum mediae و, et ى irregularium.

	I Verb. med. و	I Verb. med. و	I Verb. med. ى	IV Verb. med. و vel ى	VII Verb. med. و vel ى	VIII Verb. med. و vel ى	X Verb. med. و vel ى
Perfect. Act.	قَوَلَ	قَالَ	سَارَ	اَقَالَ	اَنْقَالَ	اِقْتَالَ	اِسْتَقَالَ
(II. P. masc. S.)	قُلْتَ	خِفْتَ	سِرْتَ	اَقَلْتَ	اَنْقَلْتَ	اِقْتَلْتَ	اِسْتَقَلْتَ
Imperf. "	يَقُولُ	يَخَافُ	يَسِيرُ	يُقِيلُ	يَنْقَالُ	يَقْتَالُ	يَسْتَقِيلُ
Imperat. "	قُلْ	خَفْ	سِرْ	اَقِلْ	اَنْقَلْ	اِقْتَلْ	اِسْتَقِلْ
Particip. "	قَائِلٌ	خَائِفٌ	سَائِرٌ	مُقِيلٌ	مُنْقَالٌ	مُقْتَالٌ	مُسْتَقِيلٌ
Perfect. Pass.	قُوِلَ	خِيفَ	سِيرَ	اُقِيلَ	اُنْقِيلَ	اُقْتِيلَ	اُسْتُقِيلَ
Imperf. "	يُقَالُ	يُخَافُ	يُسَارُ	يُقَالُ	يُنْقَالُ	يُقْتَالُ	يُسْتَقَالُ
Particip. "	مَقُولٌ	مَخُوفٌ	مَسِيرٌ	مُقَالٌ	مُنْقَالٌ	مُقْتَالٌ	مُسْتَقَالٌ
Infinitivus	قَوْلٌ	خَوْفٌ	سَيْرٌ	اِقَالَةٌ	اِنْقِيَالٌ	اِقْتِيَالٌ	اِسْتِقَالَةٌ

TABULA XV.
Paradigma flexionis
Activi I verbi ultimae و فَعَلَ

Persona	Numerus	Perfectum	Imperfectum			Imperativus
			Indicativus	Subjunctivus	Apocopatus	
3. masc.	sing.	غَزَا	يَغْزُو	يَغْزُوَ	يَغْزُ	
3. fem.	"	غَزَتْ	تَغْزُو	تَغْزُوَ	تَغْزُ	
2. masc.	"	غَزَوْتَ	تَغْزُو	تَغْزُوَ	تَغْزُ	أُغْزُ
2. fem.	"	غَزَوْتِ	تَغْزِينَ	تَغْزِي	تَغْزِي	أُغْزِي
1.	"	غَزَوْتُ	أَغْزُو	أَغْزُوَ	أَغْزُ	
3. masc.	dual.	غَزَوَا	يَغْزُوَانِ	يَغْزُوَا	يَغْزُوَا	
3. fem.	"	غَزَتَا	تَغْزُوَانِ	تَغْزُوَا	تَغْزُوَا	
2.	"	غَزَوْتُمَا	تَغْزُوَانِ	تَغْزُوَا	تَغْزُوَا	أُغْزُوَا
3. masc.	plur.	غَزَوْا	يَغْزُونَ	يَغْزُوا	يَغْزُوا	
3. fem.	"	غَزَوْنَ	يَغْزُونَ	يَغْزُونَ	يَغْزُونَ	
2. masc.	"	غَزَوْتُمْ	تَغْزُونَ	تَغْزُوا	تَغْزُوا	أُغْزُوا
2. fem.	"	غَزَوْتُنَّ	تَغْزُونَ	تَغْزُونَ	تَغْزُونَ	أُغْزُونَ
1.	"	غَزَوْنَا	نَغْزُو	نَغْزُوَ	نَغْزُ	

TABULA XVI.
Paradigma flexionis
Activi I verbi ultimae ى فَعَلَ

Persona	Numerus	Perfectum	Imperfectum Indicativus	Imperfectum Subjunctivus	Imperfectum Apocopatus	Imperativus
3. masc.	sing.	رَمَى	يَرْمِى	يَرْمِىَ	يَرْمِ	
3. fem.	„	رَمَتْ	تَرْمِى	تَرْمِىَ	تَرْمِ	
2. masc.	„	رَمَيْتَ	تَرْمِى	تَرْمِىَ	تَرْمِ	اِرْمِ
2. fem.	„	رَمَيْتِ	تَرْمِينَ	تَرْمِى	تَرْمِى	اِرْمِى
1.	„	رَمَيْتُ	أَرْمِى	أَرْمِىَ	أَرْمِ	
3. masc.	dual.	رَمَيَا	يَرْمِيَانِ	يَرْمِيَا	يَرْمِيَا	
3. fem.	„	رَمَتَا	تَرْمِيَانِ	تَرْمِيَا	تَرْمِيَا	
2.	„	رَمَيْتُمَا	تَرْمِيَانِ	تَرْمِيَا	تَرْمِيَا	اِرْمِيَا
3. masc.	plur.	رَمَوْا	يَرْمُونَ	يَرْمُوا	يَرْمُوا	
3. fem.	„	رَمَيْنَ	يَرْمِينَ	يَرْمِينَ	يَرْمِينَ	
2. masc.	„	رَمَيْتُمْ	تَرْمُونَ	تَرْمُوا	تَرْمُوا	اِرْمُوا
2. fem.	„	رَمَيْتُنَّ	تَرْمِينَ	تَرْمِينَ	تَرْمِينَ	اِرْمِينَ
1.	„	رَمَيْنَا	نَرْمِى	نَرْمِىَ	نَرْمِ	

TABULA XVII.
Paradigma flexionis
Activi I verbi ultimae ى vel فَعِلَ

Persona	Numerus	Perfectum	Imperfectum			Imperativus
			Indicativus	Subjunctivus	Apocopatus	
3. masc.	sing.	رَضِيَ	يَرْضَى	يَرْضَى	يَرْضَ	
3. fem.	"	رَضِيَتْ	تَرْضَى	تَرْضَى	تَرْضَ	
2. masc.	"	رَضِيتَ	تَرْضَى	تَرْضَى	تَرْضَ	اِرْضَ
2. fem.	"	رَضِيتِ	تَرْضَيْنَ	تَرْضَيْ	تَرْضَيْ	اِرْضَيْ
1.	"	رَضِيتُ	أَرْضَى	أَرْضَى	أَرْضَ	
3. masc.	dual.	رَضِيَا	يَرْضَيَانِ	يَرْضَيَا	يَرْضَيَا	
3. fem.	"	رَضِيَتَا	تَرْضَيَانِ	تَرْضَيَا	تَرْضَيَا	
2.	"	رَضِيتُمَا	تَرْضَيَانِ	تَرْضَيَا	تَرْضَيَا	اِرْضَيَا
3. masc.	plur.	رَضُوا	يَرْضَوْنَ	يَرْضَوْا	يَرْضَوْا	
3. fem.	"	رَضِينَ	يَرْضَيْنَ	يَرْضَيْنَ	يَرْضَيْنَ	
2. masc.	"	رَضِيتُمْ	تَرْضَوْنَ	تَرْضَوْا	تَرْضَوْا	اِرْضَوْا
2. fem.	"	رَضِيتُنَّ	تَرْضَيْنَ	تَرْضَيْنَ	تَرْضَيْنَ	اِرْضَيْنَ
1.	"	رَضِينَا	نَرْضَى	نَرْضَى	نَرْضَ	

TABULA XIX.
Paradigma stirpium verborum ultimae و et ى

	I verbi ult. و	I verbi ult. ى	I verbi ult. و vel اَ	II verbi ult. و vel ى	III verbi ult. و vel ى	IV verbi ult. و vel ى
Perfectum Activi	غَزَا	رَمَى	قَضَىَ	قَضَّى	قَاضَى	اِقْتَضَى
Imperfectum „	يَغْزُو	يَرْمِي	يَقْضَى	يُقَضِّى	يُقَاضِى	يَقْتَضِى
Imperativus „	اُغْزُ	اِرْمِ	اِقْضَ	قَضِّ	قَاضِ	اِقْتَضِ
Participium „	غَازٍ	رَامٍ	قَاضٍ	مُقَضٍّ	مُقَاضٍ	مُقْتَضٍ
Perfectum Passivi	غُزِيَ	رُمِيَ	قُضِيَ	قُضِّيَ	قُوضِيَ	اُقْتُضِيَ
Imperfectum „	يُغْزَى	يُرْمَى	يُقْضَى	يُقَضَّى	يُقَاضَى	يُقْتَضَى
Participium „	مَغْزُوٌّ	مَرْمِىٌّ	مَقْضِىٌّ	مُقَضًّى	مُقَاضًى	مُقْتَضًى
Infinitivus „	غَزْوٌ	رَمْىٌ	قَضاً		تَقْضِيَةٌ vel تَقْضاً	اِقْتِضاً

	V verbi ult. و vel ى	VI verbi ult. و vel ى	VII verbi ult. و vel ى	VIII verbi ult. و vel ى	X verbi ult. و vel ى
Perfectum Activi	تَفَتَّى	تَفَاتَى	اِنْفَتَى	اِفْتَتَى	اِسْتَفْتَى
Imperfectum "	يَتَفَتَّى	يَتَفَاتَى	يَنْفَتِى	يَفْتَتِى	يَسْتَفْتِى
Imperativus "	تَفَتَّ	تَفَاتَ	اِنْفَتِ	اِفْتَتِ	اِسْتَفْتِ
Participium "	مُتَفَتٍّ	مُتَفَاتٍ	مُنْفَتٍ	مُفْتَتٍ	مُسْتَفْتٍ
Perfectum Passivi	تُفُتِّىَ	تُفُوتِىَ	اُنْفُتِىَ	اُفْتُتِىَ	اُسْتُفْتِىَ
Imperfectum "	يُتَفَتَّى	يُتَفَاتَى	يُنْفَتَى	يُفْتَتَى	يُسْتَفْتَى
Participium "	مُتَفَتًّى	مُتَفَاتًى	مُنْفَتًى	مُفْتَتًى	مُسْتَفْتًى
Infinitivus	تَفَتٍّ	تَفَاتٍ	اِنْفِتَاءٌ	اِفْتِتَاءٌ	اِسْتِفْتَاءٌ

TABULA XVIII.
Paradigma flexionis
Passivi I verbi ultimae و vel ى

Persona	Numerus	Perfectum	Imperfectum		
			Indicativus	Subjunctivus	Apocopatus
3. masc.	sing.	قُضِيَ	يُقْضَى	يُقْضَى	يُقْضَ
3. fem.	"	قُضِيَتْ	تُقْضَى	تُقْضَى	تُقْضَ
2. masc.	"	قُضِيتَ	تُقْضَى	تُقْضَى	تُقْضَ
2. fem.	"	قُضِيتِ	تُقْضَيْنَ	تُقْضَيْ	تُقْضَيْ
1.	"	قُضِيتُ	أُقْضَى	أُقْضَى	أُقْضَ
3. masc.	dual.	قُضِيَا	يُقْضَيَانِ	يُقْضَيَا	يُقْضَيَا
3. fem.	"	قُضِيَتَا	تُقْضَيَانِ	تُقْضَيَا	تُقْضَيَا
2.	"	قُضِيتُمَا	تُقْضَيَانِ	تُقْضَيَا	تُقْضَيَا
3. masc.	plur.	قُضُوا	يُقْضَوْنَ	يُقْضَوْا	يُقْضَوْا
3. fem.	"	قُضِينَ	يُقْضَيْنَ	يُقْضَيْنَ	يُقْضَيْنَ
2. masc.	"	قُضِيتُمْ	تُقْضَوْنَ	تُقْضَوْا	تُقْضَوْا
2. fem.	"	قُضِيتُنَّ	تُقْضَيْنَ	تُقْضَيْنَ	تُقْضَيْنَ
1.	"	قُضِينَا	نُقْضَى	نُقْضَى	نُقْضَ

TABULA XX.

Paradigma flexionis nominis

a) generis masculini

α) triptoti

	indeterminati	determinati cum articulo	determinati in statu constructo
Sing. Nom.	قَصَّابٌ	اَلْقَصَّابُ	قَصَّابُ
Gen.	قَصَّابٍ	اَلْقَصَّابِ	قَصَّابِ
Acc.	قَصَّابًا	اَلْقَصَّابَ	قَصَّابَ
Dual. Nom.	قَصَّابَانِ	اَلْقَصَّابَانِ	قَصَّابَا
Gen.-Acc.	قَصَّابَيْنِ	اَلْقَصَّابَيْنِ	قَصَّابَيْ
Plur. Nom.	قَصَّابُونَ	اَلْقَصَّابُونَ	قَصَّابُو
Gen.-Acc.	قَصَّابِينَ	اَلْقَصَّابِينَ	قَصَّابِي

β) diptoti

	indeterminati	determinati cum articulo	determinati in statu constructo
Sing. Nom.	آخَرُ	اَلْآخَرُ	آخَرُ
Gen.	آخَرَ	اَلْآخَرِ	آخَرِ
Acc.	آخَرَ	اَلْآخَرَ	آخَرَ
Dual. Nom.	آخَرَانِ	اَلْآخَرَانِ	آخَرَا
Gen. Acc.	آخَرَيْنِ	اَلْآخَرَيْنِ	آخَرَيْ

	indeterminati	determinati cum articulo	determinati in statu constructo
Plur. Nom.	آخَرُونَ	اَلْآخَرُونَ	آخَرُو
Gen. Acc.	آخَرِينَ	اَلْآخَرِينَ	آخَرِي

TABULA XXI.

b) generis feminini

α) triptoti

	indeterminati	determinati cum articulo	determinati in statu constructo
Sing. Nom.	سَاعَةٌ	اَلسَّاعَةُ	سَاعَةُ
Gen.	سَاعَةٍ	اَلسَّاعَةِ	سَاعَةِ
Acc.	سَاعَةً	اَلسَّاعَةَ	سَاعَةَ
Dual. Nom.	سَاعَتَانِ	اَلسَّاعَتَانِ	سَاعَتَا
Gen.-Acc.	سَاعَتَيْنِ	اَلسَّاعَتَيْنِ	سَاعَتَيْ
Plur. Nom.	سَاعَاتٌ	اَلسَّاعَاتُ	سَاعَاتُ
Gen.-Acc.	سَاعَاتٍ	اَلسَّاعَاتِ	سَاعَاتِ

β) diptoti

Sing. Nom.	مَيَّةُ	ceterum idem
Gen. Acc.	مَيَّةَ	

PARADIGMATA.

TABULA XXII.

a) generis masculini in ـِ desinentis.

	indeterminati	determinati cum articulo	determinati in statu constructo
Sing. Nom.-Gen.	قَاضٍ	اَلْقَاضِى	قَاضِى
Acc.	قَاضِيًا	اَلْقَاضِىَ	قَاضِىَ
Dual. Nom.	قَاضِيَانِ	اَلْقَاضِيَانِ	قَاضِيَا
Gen.-Acc.	قَاضِيَيْنِ	اَلْقَاضِيَيْنِ	قَاضِيَىْ
Plur. Nom.	قَاضُونَ	اَلْقَاضُونَ	قَاضُو
Gen.-Acc.	قَاضِينَ	اَلْقَاضِينَ	قَاضِى

b) nominis in ـًى, ـًا desinentis.

α) triptoti

	indeterminati	determinati cum articulo	determinati in statu constructo
Sing. Nom.-Gen.-Acc.	مُصْطَفًى	اَلْمُصْطَفَى	مُصْطَفَى
Dual. Nom.	مُصْطَفَيَانِ	اَلْمُصْطَفَيَانِ	مُصْطَفَيَا
Gen.-Acc.	مُصْطَفَيَيْنِ	اَلْمُصْطَفَيَيْنِ	مُصْطَفَيَىْ
Plur. Nom.	مُصْطَفَوْنَ	اَلْمُصْطَفَوْنَ	مُصْطَفَوْ
Gen.-Acc.	مُصْطَفَيْنَ	اَلْمُصْطَفَيْنَ	مُصْطَفَى

	indeterminati	determinati cum articulo	determinati in statu constructo
Sing. Nom.-Gen.-Acc.	عَصًا	اَلْعَصَا	عَصَا
Dual. Nom.	عَصَوَانِ	اَلْعَصَوَانِ	عَصَوَا

β) *diptoti*

	indeterminati	determinati cum articulo	determinati in statu constructo
Sing. Nom.-Gen.-Acc.	ذِكْرَى	اَلذِّكْرَى	ذِكْرَى
id.	دُنْيَا	اَلدُّنْيَا	دُنْيَا

TABULA XXIII.
Paradigma nominis cum suffixis.

a) *nominis masc. in singulari positi* قَصَّابٌ; *fem.* جَارِيَةٌ.

cum suffixo 1. pers. sing.				قَصَّابِي	*fem.* جَارِيَتِي
"	"	2.	"	" masc.	قَصَّابُكَ
"	"	2.	"	" fem.	قَصَّابُكِ
"	"	3.	"	" masc.	قَصَّابُهُ (gen. قَصَّابِهِ)
"	"	3.	"	" fem.	قَصَّابُهَا
"	"	2.	"	dualis	قَصَّابُكُمَا
"	"	3.	"	"	قَصَّابُهُمَا (gen. قَصَّابِهِمَا)
"	"	1.	"	pluralis	قَصَّابُنَا
"	"	2.	"	" msc.	قَصَّابُكُمْ
"	"	2.	"	" fem.	قَصَّابُكُنَّ
"	"	3.	"	" msc.	قَصَّابُهُمْ (gen. قَصَّابِهِمْ)
"	"	3.	"	" fem.	قَصَّابُهُنَّ (gen. قَصَّابِهِنَّ)

b) *nominis in duali positi.*

Nominativus cum suffixo 1.pers. sing.						قَصَّابَاىَ	
”	”	”	2.	”	” msc.	قَصَّابَاكَ	etc.
Gen.-Acc.	”	”	1.	”	-	قَصَّابَىَّ	
”	”	”	2.	”	” msc.	قَصَّابَيْكَ	
”	”	”	3.	”	” ”	قَصَّابَيْهِ	
”	”	”	3.	”	” fem.	قَصَّابَيْهَا	etc.

c) *nominis masculini in plurali positi.*

Nominativus cum suffixo 1. pers. sing.						قَصَّابِىَّ	
”	”	”	2.	”	” msc.	قَصَّابُوكَ	etc.
Gen.-Acc.	”	”	1.	”	”	قَصَّابِىَّ	
”	”	”	2.	”	” msc.	قَصَّابِيكَ	
”	”	”	3.	”	” msc.	قَصَّابِيهِ	
”	”	”	3.	”	” fem.	قَصَّابِيهَا	etc.

d) *nominis feminini in plurali positi.*

Nom.-Gen.-Acc. cum suff. 1. pers. sing.						سَاعَاتِى	
Nominativus	”	”	2.	”	” msc.	سَاعَاتُكَ	
”	”	”	3.	”	” ”	سَاعَاتُهُ	etc.
Gen.-Acc.	”	”	2.	”	” ”	سَاعَاتِكَ	
”	”	”	3.	”	” ”	سَاعَاتِهِ	etc.

EXERCISES AND TEXTS.

I.

A. EXERCISES IN READING.

1. الكِتَاب مَعرِفة يقبضون نظلم ذخيرة تهتدى جميع وصلنا فريق غالب ثمين اسكنوا رجز حطط خلف بالغ هروبة شمس فرزدق بصل عام اضطرمت قال يغلظ نحرث تفلح ماض تشبه سلوك طاف ججم باع ورش حسنات درج وظيفة شاه تقنص يلحقك محابة

2. 3. قِتِلَ يُضْرَبُ دَاعٍ ظُهُورْ نَوْمْ يَكْلَبْ مِيزَانٍ صَلْوَةٌ ضَبْعَةٌ شُوَاظٌ كُبْرَى تِيهٌ رِضًى وَيْلًا مَشْرُوبٍ غِرَارَةٌ لَوْمًا يَشْتَبِهُونَ حُجَجًا مُسْتَدَانٍ سُفْلَى بُلِغُوا حَضْرَةٌ سَقَوْا حُذْيَا مَخْتُومْ تَزْدَلِعُ سُبِقُوا سُلَيْمَنْ ٭

4. إِبْلِيسْ تَأْلَفُونَ أَسَدًا مُوَنٍ أَعْبِئَةٌ أُنْزِلَ إِمْضَاءٌ مَشْنُوءَةٌ ضَأْنٌ يُطَأْطِئُ مِلْءُ هَنِيئًا هَرَأَ مُؤْتَمَرْ كَئِيبْ بَطُؤَ زَائِلْ حَمْرَاءُ يَبْدَأُ مَلْآنْ يَجِيءُ ٭

Exercises in Reading. 31*

5. وَبَّخَتِ ٱلْمُتَوَفِّيَ أَمْ يُسَيِّبُونَ ٱلنُّقَطُ ٱلشَّنِيعُ تَشَرًّا ٱلظَّالِمِينَ مَقْضِيٌّ ٱلدَّلْوُ ٱلطَّحَّانَ ٱلصِّدِّيقُ ٱلْحَيَّةَ أَحْضِرُ ٱلثَّرَاءُ ٱلْقُثَّاءَةُ ٱلذَّهَبِيُّ ٱللَّهْوُ ٱلْمُصَلَّى ٱلسَّيَّارَةُ ٱللَّهُ ٱلْقَصَّابَ ٱلْإِيَّلَ ٱلْمُفَتِّشُ ٱلتَّطْهِيرُ يُرَدُّ ٭

6. قَامَ ٱلرَّسُولُ ٭ اِعْتَزَلَ ٱلْفَرِيقَ ٭ فَٱتَّبِعُوهُ وَلَا تَتَّبِعُوا ٱلسُّبُلَ ٭ اَلضَّرْبُ ٭ اِشْهَدْ ٭ حِزْبُ ٱللَّهِ ٭ هُمُ ٱلْغَالِبُونَ ٭ عَلَى ٱلسَّطْحِ ٭ عُمَرُ بْنُ ٱلْحَرِثِ ٭ قَوْمٌ ٱفْتَرَقُوا ٭ بِسْمِ ٱلرَّحْمٰنِ ٭ زَيْدٍ ٱلطَّوِيلِ ٭ إِنِ ٱنْقَضَى ٱلْأَمْرُ ٭ لَلصَّبْرُ ٱلْجَمِيلُ ٭ عُيُونًا ٱنْبَسَطَ ٭ مَاتَتِ ٱلصَّبِيَّةُ فِي ٱلْبَيْتِ ٭ اِشْتَرَوُا ٱلثِّيرَانَ ٭ عَنِ ٱلطُّوفَانِ ٭ عَيْنَيِ ٱلْإِنْسَانِ ٭ اَلْٱفْتِرَاءُ ٭ لِلدَّالِيَةِ ٭

7. مَلْآنٌ آجِرًا تَسَآءَلُوا سَآئِقٌ قَضَاؤُهُ آذَى أَقْرِبَآءُ آنَةٌ وَزَرَآءُ أَعْدَآئِهِ يَجِيئُونَ ٭

8. 9. اَلْعَالَمِينَ يُؤْمِنُ إِلَيْهِمْ كُلُّ مَادَّةٌ هُدًى رَحْمَةٌ وَدَعَا يُفْتَرَى نُجِّيَ يَنْظُرُوا وَلِيَّةٌ ٱلسَّمٰوَاتِ ٱلرَّحْمٰنُ تَأْتِيهِمْ اِتَّبَعَنِي اِسْتَعْجَلَ ٱلْمَلَآئِكَةَ اِمْشِ تَحْمِلُهُ اِخْتَلَفُوا بَشَرٌ أَمَدَّكُمْ كَذَّبُوهُ يُوبِقْهُنَّ بَيْنَهُمْ مَسْقَطَةٌ يَشَآءُ ٭

B. EXERCISES ON THE ETYMOLOGY.

16-29. بَشَّرَ اِجْتَذَبَ تَحَارَبَ أَحْزَنَ حَسُنَ تَبَرَّعَ غَضِبَ
اِسْتَغْصَبَ اِضْطَجَعَ اِنْطَلَقَ شَاهَدَ اِخْضَرَّ اِطَّلَعَ عَمِلَ
أَقْبَلَ تَنَصَّرَ صَدَّقَ تَزَلْزَلَ اُسْتُعْمِلَ عُوَلِجَ طُحِنَ اُنْتُزِعَ
تُقُبِّلَ قُرِّبَ أُصْلِحَ تُقُوصِرَ اُضْطُرِبَ زُلْزِلَ ٭

30-33. خَبَزْتُ خَتَمْنَا أَخْرَجَتْ يَرْكَبُ أَحْسَنَ اِرْتَعَدْتُمْ تُرَصِّعُ
أُرْزَقُ يَرْجِعُونَ اِفْتَحْ حَمَلَا نَحْمِدُ اِحْتَفِظْ يُذْعَنَانِ أَدْرَكُوا
تَسْتُرْ يَرْغَبُوا نَتَكَلَّمْ اُقْعُدِى كَبُرْتُنَّ يَنْحَرِقُ أُظْهِرَتْ
نَتَحَارَبُ لُقِّبَتَا يَتَكَبَّرُ يُشَبِّهْنَ اُخْتُبِرْتِ يَنْكَشِفُ
تُقْسِمُ اِفْتَرَقُوا نَكْرَهُ نَسْتَخْرِجِينَ تَقَدَّمُوا نَاشَدْنَا
يَنْتَزِعُ أَسْنِدُوا شَرِبْتُمْ تُسَلِّطِى اِمْتَنَعْنَا اِلْبَثُوا غَسَلْتُ
تُشْرِفُونَ تُفَاخِرُ يُفْتَحَنَّ يَسْتَنْكِحُ نَتَفَقَّدْ أَقْدَرْنَا
أَعْلَمْنَ أَلْبِسُوا فَزِعْتَا تَعَرَّضْ يُسْنَدُوا اِحْتَفِرُوا نَاكِحَا

34-36. رَدَدْنَا يَضُمُّونَ أَحْبَبْتِ يَنْفَكُّ صُبَّ نَرْتَدُّ شُدِّى حَرُّوا
اِسْتَتَبَّتْ يُقْرَرْ هَمَمْتُمْ تُرَقِّيِنَ تَنْقَضُّ جُرُّوا يَعْزِزْنَ أُحِبُّوا
نُحِلَّ مُرَّ شُدِدْتَ نَسْتَقِرَّ تَصُدُّدْ ٭

EXERCISES ON THE ETYMOLOGY.

37-38. يَأْكُلُ مُرُّوا تُؤْمِنِينَ نَتَّخِذُ يَسْتَأْذِنُونَ تُوَخِّرْ اِيتَلَفْنَا
اِيلَفْ آكِلُوا بَوَّسْنَا يَسْتَنْشِرُ تَشَآءَمْتُمْ يُبْطِئُ بَطُّوْتِ
اِبْدَأْ نُسْتَبْطَأُ تَنَبَّأْتُ اِمْتَلَأْنَ تُخْطِئِينَ نُبِّئْنَا يُطَأْطِئُ ٭

40. رِدُوا يَوْسَنْ أُوسِعَتْ تَصِفْ تَرِثُوا اُسْتُولِدَ تَتَّعِدُ
نَتَوَاضَعْ يَصِلُوا تَفِدِينَ دَعُوا نَقِفْ يُوَكَّلُ تَوَجَّهْنَ
يُوجِبُ بَيَّاسْ اِسْتَيْقَظْنَا تُوَقَّظْ اِيقَظْ تَرِدْ ٭

41-44. جُزْتَ نَقُومُ أَشَرْنَا صِرْ طِرْتُمْ أَقِمْ بَاعُوا تَسْتَعِينُ
اِعْوَجَّتْ يَخْتَارُ يُمَيِّزُونَ مُتِّ خُيِّلَا كُنْ يَبِعْنَ أَبِينُوا
خِفْتُمَا اُخْتِرْنَا نَنَامُ يُرِيدَانِ تَزَوَّجْتِ أَطِيعُوا يَنْهَارُ
اِنْهَارَتْ يَسْتَعِدُّ أَطَلَّتْ غَيَّرْتُمْ يَتَصَايَحُوا نَمْ لِمْنَا تَكْ
خَافُوا اُسْتُعِينَ يُرَدّ جُدْنَا ٭

45-48. يَمْشِي أَمْسَيْتُ غَدَوْنَا أَمْضِ لَقِيَتْ نَتَعَدَّى عَمُوا
تَنْتَهِينَ يُصَلُّونَ أُنْشِي اُثْنِيَتْ تَنْجُو اِنْقَضَتْ بَكَيْتَ
بُلِينَا تُبْلَوْا يَكْفِيَانِ غَنِّ نَادَيْتُمْ يَنْبَغِي اِسْتُثْنِيَتْ
تَعْفِينَ تُبْنَى اِشْتَرُوا اُشْتَرُوا اِشْتَرُوا يُكْنَوْنَ يَتَنَحَّ
اِرْمُوا أَدْرَيْتُ رَحُوتِ اِرْضَى نَسَمّ نُودِيَتْ دَنَتْ ٭

49. يَلِي تَوْقِيتْ نَطْوِي تَشَآءَ يُؤْمُونَ اِتَّقَتْ جِنَّا يَرَوْنَ
تَرَيْنَ أَرِ أَرَيْتَ نَجِّي يَوَدُّ ⁕

53. جَعَلْنَاهُ نُثْبِتُهَا تَرْمِيهِمْ أَجَبْنَاكُمْ يَبِيعُكِ حَرَّكْتُمُوهُ
نَشْتَرِيهِ يَتْرُكُوكُمَا لُمْنِي بِجُبْنَا يَفْتَحُهَا يُعَالِجُونَنِي مَنَعُوكَ
ظَلَمْنَاهُنَّ نُحَدِّثُكُنَّ يُنَادِيهِمَا ⁕

60-61. خَادِمْ مُرْتَعِدْ مُخْرِجْ مُسَلَّطْ مَكْتُوبْ مُحْتَضَرْ مُتَحَارِبْ
مُحَرِّكْ مُجْتَمِعْ ثَاكِلْ مُتَّبِعْ مُضْطَجِعْ مُنَاكِحْ عَمَلْ مُفَاخَرَةْ
اِقْسَامْ فَتْحْ مُسْتَخْرِجْ تَحْرِيمْ تَحَارُبْ اِلْتِصَاقْ سُجُودْ
مُتَمَسِّكْ إِظْهَارْ تَكَلُّمْ نِكَاحْ تَزَعْزُعْ اِنْكِشَافْ اِسْتِقْبَالْ
مَعْمُولْ تَقْرِيبْ مُتَفَقِّدْ ⁕

67-71. سَمّْ اِنْفِكَاكْ أَذَنْ مُنْقَضْ حَاجْ اِسْتِنْبَابْ مُحِبّْ مُرْتَدّْ
مَرْفُوعْ مُسْتَقِرّْ أَعَزْ آكِلْ مَأْمُورْ إِيمَانْ مُتَّخِذْ مُؤَخَّرْ
مَمْلُوءْ قُرْآنْ اِبْطَآءْ مُمْتَلِئْ مُبْطَأْ وَسَخْ إِيسَاعْ مُنْعَدّْ
صِفَةْ اِسْتِيلَادْ مُوجِبْ مُوقِظْ وَضْعْ تَوْكِيلْ بَيْعْ خَوْفْ
مُقِيمْ صَائِرْ اِخْتِيَارْ نَائِمْ مُمَيَّزْ مُشَارْ تَصَايِحْ إِطَاعَةْ
اِسْتِعَانَةْ تَزَوُّجْ مُخَيَّلْ مُنْهَارْ مَيِّتْ مُعْوَجْ خَوْفْ
تَمْوِيتْ اِنْقِيَادْ تَغْيِيرْ مُسْتَقِيمْ سُوقْ مَقَامْ مَغِيبْ

EXERCISES ON THE SYNTAX.

88. 89.
مَشَى زَهَوْ مَقْضِيَ مُمْسٍ اِقْتِنَاءَ مُفْشًى مُتَعَدٍّ غَانٍ
مُنَادَاةٌ تَنَحْ مُسْتَثْنًى مَدْعُوٌّ اِنْبِغَاءَ جَفَاءَ رِضًى غَنِيٌّ
مُعْطَى تَسْمِيَةٌ مُنْتَهًى بَقَاءَ مُضِيَ عَدُوٌّ ۞

سُيُوفٌ أَدْيَانٌ رُؤُوسٌ رِجَالٌ عُمَّالٌ عَسَاكِرُ حِجَجٌ
أَسْمِكَةٌ غَوَاشٍ حُجَجٌ أَبْيَاتٌ مُحَفٌّ رُؤَسَاءُ أَرْجُلٌ أَغْنِيَاءُ
صُفْرٌ عَجَائِبُ عَبِيدٌ عِبَادٌ صُوَرٌ بَوَاطِنُ مَلَابِسُ أَمْكِنَةٌ
رُعَاةٌ جُهَّلٌ قَتْلَى رَعَايَا عَفَارِيتُ أَقْرِبَاءُ آلِهَةٌ سَكَارَى
أَمْطَارٌ مَمَالِكُ نَسَخَ نُجُومٌ نُوقٌ نِيرَانٌ أَيْنَامٌ أَلْسُنٌ
كَرَادِيسُ أَقْوَالٌ عَشَائِرُ سُودٌ سُودَانٌ سُرُوجٌ صِحَاحٌ
مَلَائِكَةٌ مَرَاعٍ عُمْىٌ دَوَابُّ دَرَاهِمْ رِيَاحٌ زَوَايَا سَوَاعِدُ
سَادَةٌ شُهُودٌ آثَارٌ ۞

C. EXERCISES ON THE SYNTAX.

135-138.
ضَرَبَ عَمْرٌو غُلَامًا لَهُ[1] ۞ نَزَلَ ٱلْمُؤْمِنُونَ عَلَى بَابِ
ٱلدَّارِ ۞ أُحِلَّ لَكُمْ صَيْدُ ٱلْبَحْرِ وَطَعَامُهُ ۞ كَانَتِ[2]
ٱلنِّسَاءُ فِي ٱلْجَاهِلِيَّةِ يُطَلِّقْنَ ٱلرِّجَالَ ۞ كَانَ عَبْدُ

[1] § 130. [2] § 99c.

اَللهِ عَاقِلًا مَاهِرًا¹ فِي الْعُلُومِ ٭ قَالَتِ² الْيَهُودُ وَالنَّصَارَى نَحْنُ أَبْنَاءُ اللهِ وَأَحِبَّاؤُهُ³ ٭ وَلَّى الْأَعْدَاءُ هَارِبِينَ⁴ ٭ كَانَ النَّبِيُّ يَعُودُ الْمَرِيضَ⁵ وَيَتْبَعُ الْجَنَائِزَ وَيُجَالِسُ الْفُقَرَاءَ ٭ قَالَ قَائِلٌ⁶ مِنْهُمْ⁷ لَا تَقْتُلُوا⁸ يُوسُفَ تَمَرَّغَ مُوسَى بَيْنَ يَدَيِ اللهِ تَوَاضُعًا لَهُ ٭ قَدْ جَعَلَ اللهُ قُلُوبَهُمْ قَاسِيَةً¹⁰ ٭ قَاتَلَهُمُ ابْنُ الْعَبَّاسِ¹¹ قِتَالًا شَدِيدًا¹² ٭

أَصْبَحَ¹³ النَّاسُ مِنَ النَّادِمِينَ ٭ أَمَرَ اللهُ رَسُولَهُ بِالْهِجْرَةِ وَفَرَضَ عَلَيْهِ جِهَادَ الْكُفَّارِ ٭ أُنْزِلَتِ التَّوْرِيَةُ عَلَى مُوسَى عَلَيْهِ السَّلَامُ ٭ أَقَامَ مُحَمَّدٌ بِمَكَّةَ ثَلَاثَ عَشْرَةَ سَنَةً¹⁴ ٭ أَعُوذُ بِاللهِ مِنَ الشَّيْطَانِ الرَّجِيمِ ٭ يَقْبِضُ اللهُ الْأَرْضَ يَوْمَ¹⁵ الْقِيَامَةِ وَيَطْوِي السَّمَاءَ بِيَمِينِهِ ثُمَّ يَقُولُ أَنَا الْمَلِكُ أَيْنَ¹⁶ مُلُوكُ الْأَرْضِ ٭ أَتَتْبَعُوا فِي هَذِهِ الدُّنْيَا لَعْنَةً¹⁷ ٭ كَانَ وَرَقَةُ ابْنُ¹⁸ نَوْفَلٍ قَدْ قَرَأَ¹⁹ الْكُتُبَ وَطَلَبَ الْعِلْمَ وَرَغِبَ²⁰ عَنْ

---

¹ § 110. 149. ² § 136 c 2. ³ § 124. ⁴ § 113 b. ⁵ § 118 c. ⁶ § 137 d.
⁷ § 121 a. ⁸ § 101 b. ⁹ § 113 d. ¹⁰ § 108. ¹¹ § 126. ¹² § 109. ¹³ § 110.
¹⁴ § 92 b. ¹⁵ § 113 a. ¹⁶ § 141. ¹⁷ § 108. ¹⁸ § 6 f 2. ¹⁹ § 98 e f. ²⁰ § 116.

عِبَادَةِ ٱلْأَوْثَانِ وَبَشَّرَ خَدِيجَةَ بِٱلنَّبِيِّ وَأَنَّهُ نَبِيُّ هٰذِهِ ٱلْأُمَّةِ وَأَنَّهُ سَيُؤْذَى[2] وَيُكَذَّبُ ۞ ٱلتَّوَاضُعُ سُلَّمُ ٱلشَّرَفِ ۞

139 ff. اَلْعَبِيدُ ثَلَاثَةٌ عَبْدُ رِقٍّ[3] وَعَبْدُ شَهْوَةٍ وَعَبْدُ طَمَعٍ ۞ لَهُمْ مَغْفِرَةٌ وَأَجْرٌ عَظِيمٌ ۞ لِكُلِّ شَيْءٍ رَأْسٌ وَرَأْسُ ٱلْمَعْرُوفِ تَعْجِيلُهُ ۞ قُلُوبُ ٱلْأَحْرَارِ قُبُورُ ٱلْأَسْرَارِ ۞ اَلسَّامِعُ شَرِيكُ ٱلْقَائِلِ فِي ٱلشَّرِّ ۞ اَلْأَقَارِبُ هُمُ ٱلْعَقَارِبُ ۞ اَلتَّفَكُّرُ نُورٌ وَٱلْغَفْلَةُ ظُلْمَةٌ وَٱلْجَهَالَةُ ضَلَالَةٌ وَٱلْعِلْمُ حَيَوةٌ ۞ مِنْ عَلَامَةِ ٱلْأَحْمَقِ ٱلْجُلُوسُ فَوْقَ ٱلْقَدْرِ وَٱلْمَجِيءُ فِي غَيْرِ ٱلْوَقْتِ ۞ اَلْمُلُوكُ حُكَّامٌ عَلَى ٱلنَّاسِ وَٱلْعُلَمَاءُ حُكَّامٌ عَلَى ٱلْمُلُوكِ ۞ أَحْسَنُ ٱلْكُنُوزِ مَحَبَّةُ ٱلْقُلُوبِ ۞ نَشَاطُ ٱلْمُتَكَلِّمِ بِقَدْرِ إِقْبَالِ ٱلسَّامِعِ ۞ قَالَ ٱلنَّبِيُّ ٱلْفَخْرُ فِي ٱلْإِسْلَامِ بِٱلتَّقْوَى ۞ اَلْعُذْرُ ٱلْجَمِيلُ خَيْرٌ مِنَ ٱلْمَطْلِ ٱلطَّوِيلِ ۞ إِحْدَى مَوَاجِبِ ٱلرَّحْمَةِ إِطْعَامُ ٱلْأَخِ ٱلْمُسْلِمِ ٱلْجَائِعِ ۞ اَلْبِطْنَةُ تُذْهِبُ ٱلْفِطْنَةَ ۞ حُسْنُ ٱلْخُلْقِ زِمَامٌ مِنْ رَحْمَةِ

[1] § 147 a, 148 b note. [2] § 99 a, cf. note b. [3] § 123.

اَللهِ تَعَالَى فِي أَنْفِ صَاحِبِهِ وَالزِّمَامُ بِيَدِ اَلْمَلَكِ وَاَلْمَلَكُ يَجُرُّهُ إِلَى اَلْخَيْرِ وَاَلْخَيْرُ يَجُرُّهُ إِلَى اَلْجَنَّةِ ٭ اَلْحُبُّ وَاَلْبُغْضُ يَتَوَارَثَانِ ٭ اَلصَّدِيقُ اَلْأَلُوفُ لَا يُبَاعُ بِاَلْأَلُوفِ ٭ اَلْمُنَافِقُ يُعْطِيكَ لِسَانَهُ وَيَمْنَعُكَ قَلْبَهُ ٭

147. إِنَّ اَلظَّالِمِينَ لَهُمْ عَذَابٌ أَلِيمٌ ٭ أَلَمْ تَرَ أَنَّ اَللهَ أَنْزَلَ مِنَ اَلسَّمَاءِ مَاءً فَتُصْبِحُ اَلْأَرْضُ مُخْضَرَّةً إِنَّ اَللهَ لَطِيفٌ خَبِيرٌ ٭ يَا أَيُّهَا اَلنَّبِيُّ حَرِّضِ اَلْمُؤْمِنِينَ عَلَى اَلْقِتَالِ ٭ إِنَّا أَنْزَلْنَا إِلَيْكَ اَلْكِتَابَ بِاَلْحَقِّ ٭ إِنَّ اَللهَ حَرَّمَ اَلْجَنَّةَ عَلَى اَلْمُتَكَبِّرِينَ ٭ إِنَّ اَللهَ يُحِبُّ اَلْمُحْسِنِينَ ٭ يَا اَللهُ إِنَّكَ أَنْتَ لَعَلَّامُ اَلْغُيُوبِ ٭ إِنَّ اَلْآخِرَةَ هِيَ دَارُ اَلْقَرَارِ ٭ إِنَّ اَللهَ لَسَرِيعُ اَلْحِسَابِ[1] ٭ إِنَّ اَلْمُنَافِقَ يُسِيءُ كُلَّ يَوْمٍ فَلَا[2] يَعْتَذِرُ ٭ إِنَّ فِي ذَلِكَ لَآيَاتٍ ٭ إِنَّ فِي ذَلِكَ لَعِبْرَةً لِأُولِي اَلْأَبْصَارِ ٭ إِنَّهُ لَا يُفْلِحُ اَلظَّالِمُونَ ٭ جَلَسَ اَلْإِسْكَنْدَرُ لِلنَّاسِ يَوْمًا فَلَمْ يَسْأَلْهُ أَحَدٌ حَاجَةً فَقَالَ لِجُلَسَائِهِ إِنِّي لَا أَعُدُّ هَذَا اَلْيَوْمَ مِنْ أَيَّامِ مُلْكِي ٭

[1] § 134. [2] § 152.

قَالَ بُقْرَاطُ ٱسْتَهِينُوا بِٱلْمَوْتِ فَإِنَّ مَرَارَتَهُ فِي خَوْفِهِ ۞ كُنْ فِي ٱلدُّنْيَا كَأَنَّكَ عَابِرُ سَبِيلٍ وَعُدَّ نَفْسَكَ فِي أَصْحَابِ ٱلْقُبُورِ ۞ رَزَقَكُمُ ٱللّٰهُ مِنَ ٱلطَّيِّبَاتِ لَعَلَّكُمْ تَشْكُرُونَ ۞

تَمَامُ ٱلْمُرُوءَةِ خِدْمَةُ ٱلرَّجُلِ ضَيْفَهُ[1] ۞ ٱلْقُلُوبُ أَوْعِيَةٌ وَٱلشِّفَاهُ أَقْفَالُهَا وَٱلْأَلْسُنُ مَفَاتِيحُهَا فَلْيَحْفَظْ[2] كُلُّ إِنْسَانٍ مِفْتَاحَ سِرِّهِ ۞ تَصَدَّقَ أَبُو ٱلْأَسْوَدِ عَلَى سَائِلٍ بِنَمِرَةٍ فَقَالَ[3] لَهُ جَعَلَ[4] ٱللّٰهُ نَصِيبَكَ مِنَ ٱلْجَنَّةِ مِثْلَهَا ۞ لَا تُودِعْ سِرَّكَ إِلَى طَالِبِهِ فَٱلطَّالِبُ لِلسِّرِّ[5] مُذِيعٌ ۞

مَا نَحْنُ بِتَارِكِي[6] آلِهَتِنَا ۞ ٱلْكُفَّارُ مَا هُمْ بِخَارِجِينَ مِنَ ٱلنَّارِ ۞ مَا رَبُّكَ بِظَلَّامٍ لِلْعَبِيدِ ۞ ٱلظَّالِمُونَ مَا لَهُمْ مِنْ وَلِيٍّ ۞ مَا مِنَ ٱلْأَعْمَالِ شَيْءٌ أَحَبُّ إِلَيَّ مِنْ ثَلَاثَةٍ إِشْبَاعِ جَوْعَةِ ٱلْمُسْلِمِ وَقَضَاءِ دَيْنِهِ وَتَنْفِيسِ كُرْبَتِهِ ۞ لَا يَسْتَوِي ٱلْخَبِيثُ وَٱلطَّيِّبُ ۞ قَالَ مُعَاوِيَةُ

---

[1] § 131. [2] § 101 a note. [3] § 152. [4] § 98 d. [5] § 132.
[6] § 144. [7] § 141.

كُلَّ ٱلنَّاسِ أَقْدِرُ أُرْضِيهِمْ إِلَّا¹ حَاسِدَ نِعْمَةٍ فَإِنَّهُ لَا يُرْضِيهِ إِلَّا زَوَالُهَا ✻ لَا يَتَكَبَّرُ إِلَّا كُلُّ وَضِيعٍ وَلَا يَتَوَاضَعُ إِلَّا كُلُّ رَفِيعٍ ✻ مَا نُرْسِلُ ٱلْمُرْسَلِينَ إِلَّا مُبَشِّرِينَ² وَمُنْذِرِينَ ✻

100. أُوحِىَ إِلَىَّ هٰذَا ٱلْقُرْآنُ لِأُنْذِرَكُمْ بِهِ ✻ يَنْبَغِى لِلْإِنْسَانِ أَنْ يَجْتَنِبَ مُعَاشَرَةَ ٱلْأَشْرَارِ وَيَتْرُكَ مُصَاحَبَةَ ٱلْفُجَّارِ ✻ لَا يَكُونُ ٱلصَّدِيقُ صَدِيقًا حَتَّى يَحْفَظَ أَخَاهُ فِى ثَلَاثٍ فِى نَكْبَتِهِ وَغَيْبَتِهِ وَوَفَاتِهِ ✻ نَهَى رَسُولُ ٱللَّهِ أَنْ يُنْبِعَ ٱلرَّجُلُ بَصَرَهُ لُقْمَةَ أَخِيهِ ✻ إِنَّمَا يُرِيدُ ٱلشَّيْطَانُ أَنْ يُوقِعَ بَيْنَكُمُ ٱلْعَدَاوَةَ وَٱلْبَغْضَاءَ فِى ٱلْخَمْرِ وَٱلْمَيْسِرِ وَيَصُدَّكُمْ عَنْ ذِكْرِ ٱللَّهِ وَعَنِ ٱلصَّلٰوةِ ✻

إِنَّ ٱلْعَاقِلَ يَتَّعِظُ بِٱلْأَدَبِ وَٱلْبَهَائِمُ³ لَا تَتَّعِظُ إِلَّا بِٱلضَّرْبِ ✻ قَالُوا آمَنَّا بِأَفْوَاهِهِمْ وَلَمْ تُؤْمِنْ قُلُوبُهُمْ ✻ أَوَأَمِنَ أَهْلُ ٱلْقُرَى أَنْ يَأْتِيَهُمْ بَأْسُنَا

---

¹ § 151. ² § 113 b. ³ § 157.

ضُحًى وَهُمْ يَلْعَبُونَ ۞ سَوَآءٌ عَلَيْهِمْ أَأَنْذَرْتَهُمْ أَمْ
لَمْ تُنْذِرْهُمْ ۞

154 ff. قَالُوا يَا أَمِيرَ ٱلْمُؤْمِنِينَ هِيَ ٱلدَّارُ ٱلَّتِى كُنَّا
نَسْكُنُهَا مِنْ قَبْلُ ۞ ٱلَّذِينَ آمَنُوا وَعَمِلُوا ٱلصَّالِحَاتِ
سَنُدْخِلُهُمْ جَنَّاتٍ تَجْرِى مِنْ تَحْتِهَا ٱلْأَنْهَارُ خَالِدِينَ
فِيهَا أَبَدًا ۞ قَالَ ٱللّٰهُ يَا عِيسَى ٱبْنَ مَرْيَمَ أَأَنْتَ
قُلْتَ لِلنَّاسِ ٱتَّخِذُونِى وَأُمِّى إِلٰهَيْنِ مِنْ دُونِ ٱللّٰهِ ۞
قَاتَلَ جَيْشُ ٱلْمُسْلِمِينَ ٱلْأَعْدَآءَ حَتَّى ٱنْهَزَمُوا ۞
مَثَلُ ٱلْمَلِكِ ٱلَّذِى يُعَمِّرُ خَزَائِنَهُ بِأَمْوَالِ ٱلرَّعَايَا
كَمَثَلِ ٱلَّذِى يُطَيِّنُ سَطْحَ بَيْتِهِ بِٱلتُّرَابِ ٱلَّذِى
يَقْتَلِعُهُ مِنْ أَسَاسِهِ ۞ قِيلَ لَا يُحِبُّكَ مَنْ يُحِبُّ
عَدُوَّكَ ۞ لَا تَسْتَحْقِرِ ٱلرَّأْىَ ٱلْجَلِيلَ يَأْتِيكَ بِهِ
ٱلرَّجُلُ ٱلْحَقِيرُ فَإِنَّ ٱلدُّرَّةَ ٱلْفَائِقَةَ لَا تُسْتَهَانُ لِهَوَانِ
غَائِصِهَا ۞ كَذَّبَ ٱلْقَوْمُ نُوحًا فَأَنْجَيْنَاهُ وَٱلَّذِينَ مَعَهُ
فِى ٱلْفُلْكِ وَأَغْرَقْنَا ٱلَّذِينَ كَذَّبُوا بِآيَاتِنَا ۞ ٱلصَّاحِبُ
رُقْعَةٌ فِى ٱلثَّوْبِ فَلْيَنْظُرِ ٱلْإِنْسَانُ بِمَ يَرْقَعُ ثَوْبَهُ ۞
أَدْنَى أَخْلَاقِ ٱلشَّرِيفِ كِتْمَانُ ٱلسِّرِّ وَأَعْلَى أَخْلَاقِهِ

نِسْيَانُ مَا أُسِرَّ إِلَيْهِ ۞ مَا نَدِمْتُ عَلَى مَا لَمْ أَقُلْ مَرَّةً وَنَدِمْتُ عَلَى مَا قُلْتُ مِرَارًا ۞ سَوْفَ يُنَبِّئُهُمُ ٱللَّهُ بِمَا كَانُوا يَصْنَعُونَ ۞ قِيلَ لِرَاهِبٍ مِنْ أَيْنَ تَأْكُلُ فَأَشَارَ إِلَى فِيهِ وَقَالَ ٱلَّذِى خَلَقَ هٰذِهِ ٱلرَّحَى يَأْتِيهَا بِٱلطَّحِينِ ۞ ٱلْمَحَبَّةُ شَجَرَةٌ أَصْلُهَا ٱلزِّيَارَةُ ۞

158 ff. لَا تَسْتَقِلَّ عَدُوًّا وَاحِدًا وَلَا تَسْتَكْثِرْ أَلْفَ صَدِيقٍ وَلَا تَسْتَبْدِلْ بِأَخٍ قَدِيمٍ أَخًا مُسْتَحْدَثًا مَا ٱسْتَقَامَ لَكَ ۞ اِصْحَبِ ٱلنَّاسَ كَمَا تَصْحَبُ ٱلنَّارَ خُذْ مِنْ مَنْفَعَتِهَا وَٱحْذَرْ أَنْ تُحْرِقَكَ ۞ قَالَ ٱلْإِسْكَنْدَرُ ٱنْتَفَعْتُ بِأَعْدَآئِى أَكْثَرَ مِمَّا ٱنْتَفَعْتُ بِأَصْدِقَآئِى ۞ لَقَدْ كُذِّبَتْ رُسُلٌ مِنْ قَبْلِكَ فَصَبَرُوا عَلَى مَا كُذِّبُوا وَأُوذُوا وَلَا مُبَدِّلَ لِكَلِمَاتِ ٱللَّهِ ۞

قَالَ عَلِيٌّ ٱلرِّزْقُ رِزْقَانِ رِزْقٌ تَطْلُبُهُ وَرِزْقٌ يَطْلُبُكَ فَإِنْ لَمْ تَأْتِهِ أَتَاكَ ۞ سِرُّكَ أَسِيرُكَ فَإِذَا تَكَلَّمْتَ بِهِ صِرْتَ أَسِيرَهُ ۞ ٱلْكَلَامُ كَٱلدَّوَآءِ إِنْ أَقْلَلْتَ مِنْهُ نَفَعَ وَإِنْ أَكْثَرْتَ مِنْهُ قَتَلَ ۞ نَظَرَ ٱلْإِسْكَنْدَرُ إِلَى شَيْخٍ

خَضِيب فَقَالَ لَهُ إِنْ كُنْتَ خَضَبْتَ ٱلشَّيْبَ فَكَيْفَ
تَصْنَعُ آثَارَ ٱلْكِبَرِ * لِلْمُسْلِمِ عَلَى أَخِيهِ ٱلْمُسْلِمِ مِنَ
ٱلْمَعْرُوفِ سِتٌّ يُسَلِّمُ عَلَيْهِ إِذَا لَقِيَهُ وَيَنْصَحُ لَهُ
إِذَا غَابَ عَنْهُ وَيَعُودُهُ إِذَا مَرِضَ وَيُشَيِّعُ جِنَازَتَهُ
إِذَا مَاتَ وَيُجِيبُهُ إِذَا دَعَاهُ وَيُشَمِّتُهُ إِذَا عَطَسَ *
قَالَ ٱلرَّسُولُ إِذَا أَكَلَ أَحَدُكُمْ فَلْيَأْكُلْ بِيَمِينِهِ وَإِذَا
شَرِبَ فَلْيَشْرَبْ بِيَمِينِهِ فَإِنَّ ٱلشَّيْطَانَ يَأْكُلُ بِشِمَالِهِ
وَيَشْرَبُ بِشِمَالِهِ * لِكُلِّ مَقَامٍ مَقَالٌ وَخَيْرُ ٱلْقَوْلِ
مَا وَافَقَ ٱلْحَالَ * قَالَ ٱلنَّبِيُّ إِذَا قُمْتُمْ إِلَى ٱلصَّلٰوةِ
فَٱغْسِلُوا وُجُوهَكُمْ وَأَيْدِيكُمْ إِلَى ٱلْمَرَافِقِ وَٱمْسَحُوا
بِرُؤُوسِكُمْ وَأَرْجُلِكُمْ إِلَى ٱلْكَعْبَيْنِ * ٱلْكَلِمَةُ إِذَا
خَرَجَتْ مِنَ ٱلْقَلْبِ وَقَعَتْ فِي ٱلْقَلْبِ وَإِذَا خَرَجَتْ
مِنَ ٱللِّسَانِ لَمْ تُجَاوِزِ ٱلْآذَانَ * مَنْ قَتَلَ نَفْسًا
بِغَيْرِ نَفْسٍ أَوْ فَسَادٍ فِي ٱلْأَرْضِ فَكَأَنَّمَا قَتَلَ ٱلنَّاسَ
جَمِيعًا * إِذَا دَخَلَ أَحَدُكُمُ ٱلْمَسْجِدَ فَلْيَقُلِ ٱللّٰهُمَّ
ٱفْتَحْ لِي أَبْوَابَ رَحْمَتِكَ وَإِذَا خَرَجَ فَلْيَقُلِ ٱللّٰهُمَّ
إِنِّي أَسْأَلُكَ فَضْلَكَ * إِذَا دَخَلَ رَمَضَانُ فُتِحَتْ

أَبْوَابُ ٱلسَّمَاءِ وَغُلِّقَتْ أَبْوَابُ جَهَنَّمَ وَسُلْسِلَتِ ٱلشَّيَاطِينُ ۞ تَبَاعَدُوا فِي ٱلدَّارِ تَقَارَبُوا فِي ٱلْمَوَدَّةِ ۞ اِرْحَمُوا تُرْحَمُوا وَٱغْفِرُوا يُغْفَرْ لَكُمْ ۞

إِذَا قَرَّبَكَ ٱلسُّلْطَانُ فَكُنْ مِنْهُ عَلَى حَذَرٍ وَٱحْذَرِ ٱنْقِلَابَهُ عَلَيْكَ وَكِلْمْهُ بِمَا يَشْتَهِى وَلَا يَحْمِلَنَّكَ لُطْفُهُ بِكَ عَلَى أَنْ تَدْخُلَ بَيْنَهُ وَبَيْنَ أَهْلِهِ وَحَشَمِهِ ۞ مَثَلُ ٱلْمُسْلِمِينَ فِي تَرَاحُمِهِمْ وَتَوَادِدِهِمْ وَتَوَاصُلِهِمْ كَمَثَلِ ٱلْجَسَدِ إِذَا ٱشْتَكَى عُضْوٌ مِنْهُ تَدَاعَى لَهُ سَائِرُ ٱلْجَسَدِ بِٱلْحُمَّى وَٱلسَّهَرِ ۞ إِنَّ ٱللهَ خَلَقَ ٱلْجَنَّةَ لِمَنْ أَطَاعَهُ وَلَوْ كَانَ عَبْدًا حَبَشِيًّا وَخَلَقَ ٱلنَّارَ لِمَنْ عَصَاهُ وَلَوْ كَانَ حُرًّا قُرَشِيًّا ۞ قَالَ ٱلنَّبِيُّ لَا تَرْفَعُونِي فَوْقَ قَدْرِي فَتَقُولُوا فِيَّ مَا قَالَتِ ٱلنَّصَارَى فِي ٱلْمَسِيحِ فَإِنَّ ٱللهَ عَزَّ وَجَلَّ ٱتَّخَذَنِي عَبْدًا قَبْلَ أَنْ يَتَّخِذَنِي رَسُولًا ۞ قِيلَ لِرَجُلٍ بِمَ سَادَكُمُ ٱلْأَحْنَفُ فَوَٱللهِ مَا كَانَ بِأَكْبَرِكُمْ سِنًّا وَلَا بِأَكْثَرِكُمْ مَالًا فَقَالَ بِقُوَّةِ سُلْطَانِهِ عَلَى لِسَانِهِ ۞ لَا يَقُولَنَّ أَحَدُكُمْ عَبْدِي وَأَمَتِى كُلُّكُمْ عَبِيدُ ٱللهِ وَكُلُّ نِسَائِكُمْ إِمَاءُ ٱللهِ وَلَكِنْ لِيَقُلْ

غُلَامِى وَجَارِيَتِى وَفَتَاىَ وَفَتَاتِى ۞ مَنْ حَسُنَ
خُلْقُهُ طَابَتْ عِيشَتُهُ وَدَامَتْ سَلَامَتُهُ وَتَأَكَّدَتْ فِى
ٱلنُّفُوسِ مَحَبَّتُهُ وَمَنْ سَآءَ خُلْقُهُ تَنَكَّدَتْ عِيشَتُهُ
وَدَامَتْ بِغْضَتُهُ وَنَفَرَتِ ٱلنُّفُوسُ مِنْهُ ۞ لَمَّا فَرَغَ نُوحٌ
مِنْ بِنَآءِ ٱلسَّفِينَةِ دَعَا ٱلنَّاسَ إِلَى ٱلرُّكُوبِ فِيهَا
وَأَعْلَمَهُمْ أَنَّ ٱللّٰهَ بَاعِثُ ٱلطُّوفَانِ عَلَى ٱلْأَرْضِ
كُلِّهَا حَتَّى يُطَهِّرَهَا مِنْ أَهْلِ ٱلْمَعَاصِى فَلَمْ يُجِبْهُ
أَحَدٌ مِنْهُمْ ۞ قِيلَ لِعَلِىِّ بْنِ أَبِى طَالِبٍ عَلَيْهِ
ٱلسَّلَامُ كَمْ بَيْنَ ٱلْمَشْرِقِ وَٱلْمَغْرِبِ قَالَ مَسِيرَةُ يَوْمٍ
لِلشَّمْسِ قِيلَ لَهُ كَمْ بَيْنَ ٱلسَّمَآءِ وَٱلْأَرْضِ قَالَ
مَسِيرَةُ سَاعَةٍ لِدَعْوَةٍ مُسْتَجَابَةٍ ۞ قِيلَ لِبَعْضِ
ٱلْكُرَمَآءِ كَيْفَ ٱكْتَسَبْتَ مَكَارِمَ ٱلْأَخْلَاقِ وَٱلتَّأَدُّبَ
مَعَ ٱلْأَضْيَافِ فَقَالَ كَانَتِ ٱلْأَسْفَارُ تُحْوِجُنِى إِلَى أَنْ
أَفِدَ عَلَى ٱلنَّاسِ فَمَا ٱسْتَحْسَنْتُ مِنْ أَخْلَاقِهِمْ ٱتَّبَعْتُهُ
وَمَا ٱسْتَقْبَحْتُهُ ٱجْتَنَبْتُهُ ۞ حَضَرَ أَعْرَابِىٌّ عَلَى مَائِدَةِ
بَعْضِ ٱلْخُلَفَآءِ فَقُدِّمَ جَدْىٌ مَشْوِىٌّ فَجَعَلَ ٱلْأَعْرَابِىُّ
يُسْرِعُ فِى أَكْلِهِ مِنْهُ فَقَالَ لَهُ ٱلْخَلِيفَةُ تَأْكُلُهُ بِحَرْدٍ

كَأَنَّ أُمَّهُ نَطَحَتْكَ فَقَالَ أَرَاكَ تُشْفِقُ عَلَيْهِ كَأَنَّ أُمَّهُ أَرْضَعَتْكَ ٭

كَانَ مِنْ سُنَنِ ٱلْعَرَبِ نِكَاحُ ٱلْمَقْتِ وَهُوَ أَنَّ ٱلرَّجُلَ إِذَا مَاتَ قَامَ وَلَدُهُ ٱلْأَكْبَرُ فَأَلْقَى ثَوْبَهُ عَلَى ٱمْرَأَةِ أَبِيهِ فَوَرِثَ نِكَاحَهَا فَإِنْ لَمْ يَكُنْ لَهُ بِهَا حَاجَةٌ زَوَّجَهَا لِبَعْضِ إِخْوَتِهِ بِمَهْرٍ جَدِيدٍ فَكَانُوا يَرِثُونَ ٱلنِّكَاحَ كَمَا يَرِثُونَ ٱلْمَالَ ٭ كَانَ عَبْدُ ٱللّٰهِ ٱبْنُ مَرْزُوقٍ مِنْ نُدَمَاءِ ٱلْمَهْدِيِّ فَسَكِرَ يَوْمًا فَفَاتَتْهُ ٱلصَّلَاةُ فَجَآءَتْهُ جَارِيَةٌ لَهُ بِجَمْرَةٍ فَوَضَعَتْهَا عَلَى رِجْلِهِ فَٱنْتَبَهَ مَذْعُورًا فَقَالَتْ لَهُ إِذَا لَمْ تَصْبِرْ عَلَى نَارِ ٱلدُّنْيَا فَكَيْفَ تَصْبِرُ عَلَى نَارِ ٱلْآخِرَةِ ٭ دَخَلَ عَامِلٌ لِعُمَرَ بْنِ ٱلْخَطَّابِ رَضِيَ ٱللّٰهُ عَنْهُ فَوَجَدَهُ مُسْتَلْقِيًا عَلَى ظَهْرِهِ وَصِبْيَانُهُ يَلْعَبُونَ عَلَى بَطْنِهِ فَأَنْكَرَ ذٰلِكَ عَلَيْهِ فَقَالَ لَهُ عُمَرُ كَيْفَ أَنْتَ مَعَ أَهْلِكَ قَالَ إِذَا دَخَلْتُ سَكَتَ ٱلنَّاطِقُ فَقَالَ لَهُ ٱعْتَزِلْ فَإِنَّكَ لَا تَرْفِقُ بِأَهْلِكَ وَوَلَدِكَ فَكَيْفَ تَرْفِقُ بِأُمَّةِ مُحَمَّدٍ صَلَّى ٱللّٰهُ عَلَيْهِ وَسَلَّمَ ٭

قَالَ بَعْضُ ٱلْمُلُوكِ لِبَعْضِ وُزَرَآئِهِ وَأَرَادَ جِنَّتَهُ مَا
خَيْرُ مَا يُرْزَقُهُ ٱلْعَبْدُ قَالَ عَقْلٌ يَعِيشُ بِهِ قَالَ فَإِنْ
عَدِمَهُ قَالَ فَأَدَبٌ يَتَحَلَّى بِهِ قَالَ فَإِنْ عَدِمَهُ قَالَ
فَمَالٌ يَسْتُرُهُ قَالَ فَإِنْ عَدِمَهُ قَالَ فَصَاعِقَةٌ تُحْرِقُهُ
فَتُرِيحُ مِنْهُ ٱلْعِبَادَ وَٱلْبِلَادَ ٭

نَزَلَ رَجُلٌ بِصَوْمَعَةِ رَاهِبٍ فَقَدَّمَ إِلَيْهِ ٱلرَّاهِبُ أَرْبَعَةَ
أَرْغِفَةٍ وَذَهَبَ لِيُحْضِرَ إِلَيْهِ ٱلْعَدَسَ فَحَمَلَهُ وَجَآءَ
فَوَجَدَهُ قَدْ أَكَلَ ٱلْخُبْزَ فَذَهَبَ فَأَتَى بِخُبْزٍ فَوَجَدَهُ قَدْ
أَكَلَ ٱلْعَدَسَ فَفَعَلَ مَعَهُ ذَلِكَ عَشَرَ مَرَّاتٍ فَسَأَلَهُ
ٱلرَّاهِبُ أَيْنَ مَقْصِدُهُ قَالَ إِلَى ٱلْأُرْدُنِّ قَالَ لِمَا ذَا قَالَ
بَلَغَنِى أَنَّ بِهَا طَبِيبًا حَاذِقًا أَسْأَلُهُ عَمَّا يُصْلِحُ
مَعِدَتِى فَإِنِّى قَلِيلُ ٱلشَّهْوَةِ لِلطَّعَامِ فَقَالَ لَهُ ٱلرَّاهِبُ
إِنَّ لِى إِلَيْكَ حَاجَةً قَالَ وَمَا هِىَ قَالَ إِذَا ذَهَبْتَ
وَأَصْلَحْتَ مَعِدَتَكَ فَلَا تَجْعَلْ رُجُوعَكَ عَلَىَّ ٭

II.
ARABIC PROSE EXTRACTS.

زَعَمُوا أَنَّ مَلِكًا يُقَالُ لَهُ ٱلْهُدْهَادُ خَرَجَ لِلصَّيْدِ
فِى جَمَاعَةٍ مِنْ خَدَمِهِ وَخَاصَّتِهِ فَرَأَى غَزَالًا يَطْرُدُهُ
ذِئْبٌ وَقَدْ أَضَاقَتْهُ إِلَى ضِيقٍ لَيْسَ لِلْغَزَالِ مِنْهُ مَخْلَصٌ
فَحَمَلَ ٱلْهُدْهَادُ عَلَى ٱلذِّئْبِ حَتَّى طَرَدَهُ عَنِ ٱلْغَزَالِ
وَخَلَّصَ ٱلْغَزَالَ مِنْهُ فَسَارَ فِى أَثَرِ ٱلْغَزَالِ وَٱنْقَطَعَ عَنْهُ
أَصْحَابُهُ فَبَيْنَا هُوَ كَذَلِكَ إِذْ ظَهَرَتْ لَهُ مَدِينَةٌ عَظِيمَةٌ
فِيهَا مِنْ كُلِّ شَىْءٍ مِنَ ٱلشَّاءِ وَٱلنَّعَمِ وَٱلنَّخْلِ
وَٱلزَّرْعِ وَأَنْوَاعِ ٱلْفَوَاكِهِ فَوَقَفَ دُونَهَا مُتَعَجِّبًا مِمَّا
ظَهَرَ لَهُ إِذْ أَقْبَلَ رَجُلٌ مِنْ أَهْلِ تِلْكَ ٱلْمَدِينَةِ ٱلَّتِى
ظَهَرَتْ لَهُ فَسَلَّمَ وَرَحَّبَ بِهِ ثُمَّ قَالَ لَهُ أَيُّهَا ٱلْمَلِكُ
إِنِّى أَرَاكَ مُتَعَجِّبًا مِمَّا ظَهَرَ لَكَ فِى يَوْمِكَ هَذَا فَقَالَ
لَهُ ٱلْهُدْهَادُ إِنِّى لَكَمَا قُلْتَ فَمَا هَذِهِ ٱلْمَدِينَةُ وَمَنْ
سَاكِنُهَا فَقَالَ هَذِهِ مَدِينَةُ مَأْرِبَ وَسُكَّانُهَا حَىٌّ مِنَ
ٱلْجِنِّ وَأَنَا مَلِكُهُمْ وَصَاحِبُ أَمْرِهِمْ فَهُوَ مَعَهُ فِى هَذَا
ٱلْكَلَامِ إِذْ مَرَّتْ بِهِمَا ٱمْرَأَةٌ لَمْ يَرَ ٱلرَّاوُونَ أَحْسَنَ

مِنْهَا وَجْهًا وَلَا أَكْمَلَ مِنْهَا خَلْقًا وَلَا أَظْهَرَ مِنْهَا
صَبَاحَةً وَلَا أَطْيَبَ رَائِحَةً فَافْتُتِنَ بِهَا ٱلْهُدْهُدُ وَعَلِمَ
مَلِكُ ٱلْجِنِّ أَنَّهُ قَدْ هَوِيَهَا وَشَغِفَ بِهَا فَقَالَ إِنْ كُنْتَ
قَدْ هَوِيتَهَا فَهِيَ ٱبْنَتِى فَأَنَا أُزَوِّجُكَهَا فَجَازَاهُ ٱلْهُدْهُدُ
خَيْرًا فَقَالَ لَهُ ٱلْجِنِّىُّ هَلْ عَرَفْتَهَا قَالَ ٱلْهُدْهُدُ مَا
رَأَيْتُهَا قَبْلَ يَوْمِى هٰذَا قَالَ ٱلْجِنِّىُّ هِىَ ٱلْغَزَالُ ٱلَّتِى
خَلَّصْتَهَا مِنَ ٱلذِّئْبِ وَلَا نُكَافِئُكَ عَلَى جَمِيلٍ فِعْلِكَ
أَبَدًا بِأَحْسَنَ مِنْ أَنْ نَخْبُوَكَ بِهَا فَتَأَهَّبْ لِدُخُولِكَ
عَلَيْهَا فَإِنِّى قَدْ زَوَّجْتُكَهَا بِشَهَادَةِ ٱللّٰهِ تَعَالَى وَشَهَادَةِ
مَلَائِكَتِهِ فَإِذَا أَرَدْتَ ذٰلِكَ فَأَقْدِمْ إِلَيْنَا بِخَاصَّةِ أَهْلِ
بَيْتِكَ وَمُلُوكِ قَوْمِكَ لِيَشْهَدُوا مَلَاكَهَا وَيَحْضُرُوا وَلِيمَتَهَا
وَمِيعَادُكَ ٱلشَّهْرُ ٱلدَّاخِلُ فَٱنْصَرَفَ ٱلْهُدْهُدُ عَلَى
ٱلْمِيعَادِ وَغَابَتِ ٱلْمَدِينَةُ عَنْهُ فَإِذَا أَصْحَابُهُ حَوْلَهُ
يَدُورُونَ لَهُ فَقَالُوا أَيْنَ كُنْتَ فَنَحْنُ فِى طَلَبِكَ مُنْذُ
فَارَقْتَنَا وَلَمْ نَتْرُكْ شَيْئًا مِنْ هٰذِهِ ٱلْفَلَوَاتِ إِلَّا طَلَبْنَاكَ
فِيهِ فَزَعَمُوا أَنَّ ٱلْهُدْهُدَ خَرَجَ عَلَى ٱلْمِيعَادِ إِلَى
أَصْهَارِهِ مِنَ ٱلْجِنِّ فِى خَاصَّةِ قَوْمِهِ وَخَدَمِهِ حَتَّى وَافَاهُمْ

فَوَجَدُوا قَصْرًا بَنَاهُ لَهُ ٱلْجِنُّ فِي فَلَاةٍ مِنَ ٱلْأَرْضِ تَحْفُوهُ بِٱلنَّخْلِ وَٱلْأَعْنَابِ وَأَلْوَانِ ٱلزَّرْعِ وَأَنْوَاعِ ٱلْفَوَاكِهِ تَخْتَرِقُ فِيهَا ٱلْمِيَاهُ ٱلْجَارِيَةُ فَتَعَجَّبَ ٱلْقَوْمُ مِنْ ذٰلِكَ تَعَجُّبًا شَدِيدًا وَرَأَوْا مَلِكًا عَظِيمًا فَنَزَلُوا فِي ٱلْقَصْرِ مَعَهُ عَلَى فُرُشٍ لَمْ يَرَوْا مِثْلَهَا وَثُرِّبَتْ لَهُمْ مَوَائِدُ عَلَيْهَا مِنْ طَيِّبَاتِ ٱلْمَأْكُولِ وَأَلْوَانِهَا ٱلَّتِي لَمْ يَأْكُلُوا قَطُّ أَطْيَبَ مِنْهَا طَعَامًا وَلَا أَزْكَى مِنْهَا رَائِحَةً وَشَرِبُوا مِنَ ٱلشَّرَابِ مَا لَمْ يَشْرَبُوا قَطُّ أَلَذَّ وَلَا أَخَفَّ مِنْهُ فَمَكَثُوا مَعَهُ ثَلَاثَةَ أَيَّامٍ بِلَيَالِيهَا فِي ذٰلِكَ وَزُفَّتْ إِلَى ٱلْهُدْهَادِ ٱمْرَأَتُهُ بِنْتُ مَلِكِ ٱلْجِنِّ وَأَذِنَ ٱلْهُدْهَادُ لِبَنِي عَمِّهِ وَخَاصَّتِهِ وَعَشِيرَتِهِ بِٱلِٱنْصِرَافِ إِلَى مَوَاضِعِهِمْ وَصَارَ ذٰلِكَ ٱلْقَصْرُ دَارَ مَمْلَكَتِهِ فَزَعَمُوا أَنَّهُ مَكَثَ زَمَانًا مَعَ ٱمْرَأَتِهِ وَأَوْلَدَهَا بِلْقِيسَ ✻

أَجْمَعَتْ قُرَيْشٌ عَلَى قَتْلِ رَسُولِ ٱللّٰهِ وَقَالُوا لَيْسَ لَهُ ٱلْيَوْمَ أَحَدٌ يَنْصُرُهُ وَقَدْ مَاتَ أَبُو طَالِبٍ فَأَجْمَعُوا جَمِيعًا أَنْ يَأْتُوا مِنْ كُلِّ قَبِيلَةٍ بِغُلَامٍ نَهْدٍ فَيَجْتَمِعُوا عَلَيْهِ فَيَضْرِبُوهُ بِأَسْيَافِهِمْ ضَرْبَةَ رَجُلٍ وَاحِدٍ فَلَا يَكُونَ

لِبَنِى هَاشِمٍ قُوَّةٌ بِمُعَادَاةِ جَمِيعِ قُرَيْشٍ فَبَلَغَ رَسُولَ ٱللّٰهِ ذٰلِكَ وَلَمَّا ٱخْتَلَطَ ٱلظَّلَامُ خَرَجَ وَمَعَهُ أَبُو بَكْرٍ وَخَلَّفَ عَلِيًّا عَلَى فِرَاشِهِ لِرَدِّ ٱلْوَدَائِعِ ٱلَّتِى كَانَتْ عِنْدَهُ وَصَارَ إِلَى ٱلْغَارِ ٱلَّذِى كَانَ يَتَحَنَّثُ فِيهِ قَبْلَ ٱلنُّبُوَّةِ فَكَمِنَ فِيهِ وَأَتَتْ قُرَيْشٌ فِرَاشَهُ فَوَجَدُوا عَلِيًّا فَقَالُوا أَيْنَ ٱبْنُ عَمِّكَ قَالَ قُلْتُمْ لَهُ ٱخْرُجْ عَنَّا فَخَرَجَ عَنْكُمْ فَطَلَبُوا ٱلْأَثَرَ فَلَمْ يَقَعُوا عَلَيْهِ وَأَعْمَى ٱللّٰهُ عَلَيْهِمُ ٱلْمَوَاضِعَ فَوَقَفُوا عَلَى بَابِ ٱلْغَارِ وَقَدْ عَشَّشَتْ عَلَيْهِ حَمَامَةٌ فَقَالُوا مَا فِى هٰذَا ٱلْغَارِ أَحَدٌ وَٱنْصَرَفُوا فَخَرَجَ رَسُولُ ٱللّٰهِ مُتَوَجِّهًا إِلَى ٱلْمَدِينَةِ فَعَلِمَتْ قُرَيْشٌ أَنَّهُ قَدْ مَضَى إِلَى يَثْرِبَ وَٱتَّبَعَهُ سُرَاقَةُ فَلَمَّا لَحِقَهُ قَالَ رَسُولُ ٱللّٰهِ ٱللّٰهُمَّ ٱكْفِنَا سُرَاقَةَ فَسَاخَتْ قَوَائِمُ فَرَسِهِ فَصَاحَ يَاٱبْنَ أَبِى قُحَافَةَ قُلْ لِصَاحِبِكَ أَنْ يَدْعُوَ ٱللّٰهَ بِإِطْلَاقِ فَرَسِى فَلَعَمْرِى لَئِنْ لَمْ يُصِبْهُ مِنِّى خَيْرٌ فَلَا يُصِبْهُ مِنِّى شَرٌّ فَلَمَّا رَجَعَ إِلَى مَكَّةَ خَبَّرَهُمُ ٱلْخَبَرَ فَكَذَّبُوهُ ❈

حُكِىَ عَنِ ٱلرَّبِيعِ حَاجِبِ ٱلْخَلِيفَةِ ٱلْمَنْصُورِ قَالَ مَا رَأَيْتُ رَجُلًا أَرْبَطَ جَأْشًا وَأَثْبَتَ جَنَانًا مِنْ رَجُلٍ

سَعَى بِهِ إِلَى ٱلْمَنْصُورِ أَنَّ عِنْدَهُ وَدَائِعَ وَأَمْوَالًا لِبَنِى
أُمَيَّةَ فَأَمَرَنِى بِإِحْضَارِهِ فَأَحْضَرْتُهُ إِلَيْهِ فَقَالَ لَهُ ٱلْمَنْصُورُ
قَدْ رُفِعَ إِلَيْنَا خَبَرُ ٱلْوَدَائِعِ وَٱلْأَمْوَالِ ٱلَّتِى عِنْدَكَ لِبَنِى
أُمَيَّةَ فَأَخْرِجْهَا لَنَا وَلَا تَكْتُمْ مِنْهَا شَيْئًا فَقَالَ يَا أَمِيرَ
ٱلْمُؤْمِنِينَ أَأَنْتَ وَارِثٌ لِبَنِى أُمَيَّةَ قَالَ لَا قَالَ فَوَصَّى
لَهُمْ فِى أَمْوَالِهِمْ وَرِبَاعِهِمْ قَالَ لَا قَالَ فَمَا مَسْأَلَتُكَ
عَمَّا فِى يَدِى مِنْ ذَلِكَ قَالَ فَأَطْرَقَ ٱلْمَنْصُورُ وَتَفَكَّرَ
سَاعَةً ثُمَّ رَفَعَ رَأْسَهُ وَقَالَ إِنَّ بَنِى أُمَيَّةَ ظَلَمُوا ٱلْمُسْلِمِينَ
فِيهَا وَأَنَا وَكِيلُ ٱلْمُسْلِمِينَ فِى حُقُوقِهِمْ وَأُرِيدُ أَنْ آخُذَ
مَا ظَلَمُوا ٱلْمُسْلِمِينَ فِيهِ فَأَجْعَلَهُ فِى بَيْتِ أَمْوَالِهِمْ
فَقَالَ يَا أَمِيرَ ٱلْمُؤْمِنِينَ فَتَحْتَاجُ إِلَى إِقَامَةِ بَيِّنَةٍ عَادِلَةٍ
أَنَّ مَا فِى يَدِى لِبَنِى أُمَيَّةَ هُوَ ٱلَّذِى غَصَبُوهُ مِنَ ٱلنَّاسِ
فَإِنَّ بَنِى أُمَيَّةَ قَدْ كَانَتْ لَهُمْ أَمْوَالٌ غَيْرُ أَمْوَالِ
ٱلْمُسْلِمِينَ قَالَ فَأَطْرَقَ ٱلْمَنْصُورُ سَاعَةً ثُمَّ رَفَعَ رَأْسَهُ
وَقَالَ يَا رَبِيعُ مَا أَرَى ٱلشَّيْخَ إِلَّا قَدْ صَدَقَ وَمَا
يَجِبُ عَلَيْهِ شَىْءٌ وَمَا يَسَعُنَا إِلَّا أَنْ نَعْفُوَ عَمَّا قِيلَ
عَنْهُ ثُمَّ قَالَ هَلْ لَكَ مِنْ حَاجَةٍ قَالَ نَعَمْ حَاجَتِى يَا

أَمِيرَ ٱلْمُؤْمِنِينَ أَنْ تَجْمَعَ بَيْنِى وَبَيْنَ مَنْ سَعَى فِى
إِلَيْكَ فَوَٱللهِ ٱلَّذِى لَا إِلٰهَ إِلَّا هُوَ مَا فِى يَدَىَّ لِبَنِى
أُمَيَّةَ مَالٌ وَلَا وَدِيعَةٌ فَقَالَ ٱلْخَلِيفَةُ يَا رَبِيعُ ٱجْمَعْ
بَيْنَهُ وَبَيْنَ مَنْ سَعَى بِهِ فَجَمَعْتُ بَيْنَهُمَا فَلَمَّا رَآهُ
قَالَ هٰذَا غُلَامِى ٱخْتَلَسَ لِى ثَلَاثَةَ آلَافِ دِينَارٍ مِنْ
مَالِى وَأَبَقَ مِنِّى وَخَافَ مِنْ طَلَبِى لَهُ فَسَعَى بِى عِنْدَ
أَمِيرِ ٱلْمُؤْمِنِينَ قَالَ فَشَدَّدَ ٱلْمَنْصُورُ عَلَى ٱلْغُلَامِ وَخَوَّفَهُ
فَأَقَرَّ بِأَنَّهُ غُلَامُهُ وَأَنَّهُ أَخَذَ ٱلْمَالَ ٱلَّذِى ذَكَرَهُ وَسَعَى
بِهِ كِذْبًا عَلَيْهِ وَخَوْفًا مِنْ أَنْ يَقَعَ فِى يَدِهِ فَقَالَ لَهُ
ٱلْمَنْصُورُ سَأَلْتُكَ أَيُّهَا ٱلشَّيْخُ أَنْ تَعْفُوَ عَنْهُ فَقَالَ قَدْ
عَفَوْتُ عَنْهُ وَأَعْتَقْتُهُ وَوَهَبْتُهُ ثَلَاثَةَ ٱلْآلَافِ ٱلَّتِى أَخَذَهَا
وَثَلَاثَةَ آلَافٍ أُخْرَى أَدْفَعُهَا إِلَيْهِ فَقَالَ لَهُ ٱلْمَنْصُورُ
مَا عَلَى مَا فَعَلْتَ مِنْ مَزِيدٍ قَالَ بَلَى يَا أَمِيرَ ٱلْمُؤْمِنِينَ
إِنَّ هٰذَا كُلَّهُ لَقَلِيلٌ فِى مُقَابَلَةِ كَلَامِكَ لِى وَعَفْوِكَ
عَنِّى ثُمَّ ٱنْصَرَفَ قَالَ ٱلرَّبِيعُ فَكَانَ ٱلْمَنْصُورُ يَتَعَجَّبُ
مِنْهُ وَكُلَّمَا ذَكَرَهُ يَقُولُ مَا رَأَيْتُ مِثْلَ هٰذَا ٱلشَّيْخِ
يَا رَبِيعُ ٭

رُوِيَ أَنَّ ٱلْإِسْكَنْدَرَ وَهُوَ ٱلَّذِي يُقَالُ لَهُ ذُو ٱلْقَرْنَيْنِ مَلَكَ بَعْدَ أَبِيهِ وَكَانَ مُعَلِّمَهُ أَرَسْطَاطَالِيسُ ٱلْحَكِيمُ فَجَلَّ قَدْرُ ٱلْإِسْكَنْدَرِ وَعَظُمَ مُلْكُهُ وَٱشْتَدَّ سُلْطَانُهُ وَأَعَانَتْهُ ٱلْحِكْمَةُ وَٱلْعَقْلُ وَٱلْمَعْرِفَةُ وَكَانَ مَعَهُ نَجْدَةٌ وَبَأْسٌ وَهِمَّةٌ عَالِيَةٌ دَعَتْهُ إِلَى أَنْ كَتَبَ إِلَى مُلُوكِ ٱلْأَقَالِيمِ وَٱلْآفَاقِ يَدْعُوهُمْ إِلَى طَاعَتِهِ وَمَنْ كَانَ قَبْلَهُ مِنْ مُلُوكِ ٱلْيُونَانِيِّينَ يُؤَدِّي إِلَى مُلُوكِ أَرْضِ بَابِلَ مِنَ ٱلْفُرْسِ خَرْجًا لِجَلَالَةِ تِلْكَ ٱلْمَمْلَكَةِ وَعِظَمِ قَدْرِهَا وَصِغَرِ ٱلْمَمَالِكِ فِي جَنْبِهَا فَلَمَّا كَتَبَ إِلَى مَلِكِ فَارِسَ يَدْعُوهُ إِلَى طَاعَتِهِ عَظُمَ عَلَيْهِ فَسَارَ ٱلْإِسْكَنْدَرُ حَتَّى أَتَى أَرْضَ بَابِلَ وَمَلِكُ ٱلْفُرْسِ يَوْمَئِذٍ دَارَا فَحَارَبَهُ حَتَّى قَتَلَهُ وَحَوَى خَزَائِنَ مُلْكِهِ وَتَزَوَّجَ ٱبْنَتَهُ ثُمَّ صَارَ إِلَى أَرْضِ فَارِسَ وَقَتَلَ بِهَا مِنَ ٱلْمَرَازِبَةِ وَٱلرُّوسَاءِ وَٱفْتَتَحَ ٱلْبِلَادَ ثُمَّ صَارَ إِلَى أَرْضِ ٱلْهِنْدِ فَزَحَفَ إِلَيْهِ مَلِكُ ٱلْهِنْدِ فَحَارَبَهُ حَتَّى قَتَلَهُ ثُمَّ صَيَّرَ ٱلْإِسْكَنْدَرُ عَلَى ٱلْهِنْدِ مَلِكًا مِنْ قِبَلِهِ مِنْ أَهْلِ ٱلْهِنْدِ ثُمَّ رَجَعَ إِلَى أَرْضِ بَابِلَ بَعْدَ أَنْ دَوَّخَ

الْأَرْضَ فَلَمَّا صَارَ فِي أَدَانِي ٱلْعِرَاقِ مِمَّا يَلِي ٱلْجَزِيرَةَ
ٱعْتَلَّ فَٱشْتَدَّتْ عِلَّتُهُ فَلَمَّا يَئِسَ مِنْ نَفْسِهِ وَعَلِمَ
أَنَّ ٱلْمَوْتَ قَدْ نَزَلَ بِهِ كَتَبَ إِلَى أُمِّهِ كِتَابًا يُعَزِّيهَا
عَنْ نَفْسِهِ وَقَالَ لَهَا فِي آخِرِهِ ٱصْنَعِي طَعَامًا وَٱجْمَعِي
مَنْ قَدَرْتِ عَلَيْهِ مِنْ نِسَاءِ ٱلْمَمْلَكَةِ وَلَا يَأْكُلْ مِنْ
طَعَامِكِ مَنْ أُصِيبَ بِمُصِيبَةٍ قَطُّ فَعَمِلَتْ طَعَامًا
وَجَمَعَتِ ٱلنَّاسَ ثُمَّ أَمَرَتْهُمْ أَلَّا يَأْكُلَ مَنْ أُصِيبَ
بِمُصِيبَةٍ قَطُّ فَلَمْ يَأْكُلْ أَحَدٌ فَعَلِمَتْ مَا أَرَادَ وَمَاتَ
ٱلْإِسْكَنْدَرُ فِي مَوْضِعِهِ ٱلَّذِي كَاتَبَ مِنْهُ فَٱجْتَمَعَ أَصْحَابُهُ
فَكَفَّنُوهُ وَحَنَّطُوهُ وَصَيَّرُوهُ فِي تَابُوتٍ مِنْ ذَهَبٍ ثُمَّ وَقَفَ
عَلَيْهِ عَظِيمٌ مِنَ ٱلْفَلَاسِفَةِ فَقَالَ هَذَا يَوْمٌ عَظِيمٌ
أَقْبَلَ مِنْ شَرِّهِ مَا كَانَ مُدْبِرًا وَأَدْبَرَ مِنْ خَيْرِهِ مَا
كَانَ مُقْبِلًا ثُمَّ أَقْبَلَ عَلَى مَنْ حَضَرَهُ مِنَ ٱلْفَلَاسِفَةِ
فَقَالَ يَا مَعْشَرَ ٱلْحُكَمَاءِ لِيَقُلْ كُلُّ ٱمْرِئٍ مِنْكُمْ قَوْلًا
يَكُونُ لِلْخَاصَّةِ مُعَزِّيًا وَلِلْعَامَّةِ وَاعِظًا فَقَامَ كُلُّ وَاحِدٍ
مِنْ تَلَامِذَةِ أَرَسْطَاطَالِيسَ فَضَرَبَ بِيَدِهِ عَلَى ٱلتَّابُوتِ
ثُمَّ قَالَ أَيُّهَا ٱلْمَنْطِيقِ مَا أَخْرَسَكَ أَيُّهَا ٱلْعَزِيزُ

مَا أَذَلَّكَ أَيُّهَا ٱلْقَانِصُ أَنَّى وَقَعْتَ مَوْضِعَ ٱلصَّيْدِ فِي شَرَكِ ٱلَّذِي يَقْنِصُكَ ثُمَّ قَامَ آخَرُ فَقَالَ هٰذَا ٱلْقَوِيُّ ٱلَّذِي أَصْبَحَ ٱلْيَوْمَ ضَعِيفًا وَٱلْعَزِيزُ ٱلَّذِي أَصْبَحَ ٱلْيَوْمَ ذَلِيلًا وَقَامَ آخَرُ فَقَالَ قَدْ كَانَتْ سُيُوفُكَ لَا تَنْجَفُّ وَنِقَمَاتُكَ لَا تُؤْمَنُ وَكَانَتْ مَدَائِنُكَ لَا تُرَامُ وَكَانَتْ عَطَايَاكَ لَا تَبْرَحُ وَكَانَ ضِيَاؤُكَ لَا يُكْسَفُ فَأَصْبَحَ ضَوْءُكَ قَدْ خَمَدَ وَنِقَمَاتُكَ لَا تُخْشَى وَأَصْبَحَتْ عَطَايَاكَ لَا تُرْجَى وَأَصْبَحَتْ سُيُوفُكَ لَا تُنْتَضَى وَأَصْبَحَتْ مَدَائِنُكَ لَا تُمْنَعُ وَقَامَ آخَرُ فَقَالَ قَدْ كَانَ صَوْتُكَ مَرْهُوبًا وَكَانَ مُلْكُكَ غَالِبًا فَأَصْبَحَ ٱلصَّوْتُ قَدِ ٱنْقَطَعَ وَٱلْمُلْكُ قَدِ ٱتَّضَعَ وَقَامَ آخَرُ فَقَالَ حَرَّكَنَا ٱلْإِسْكَنْدَرُ بِسُكُونِهِ وَأَنْطَقَنَا بِصُمُوتِهِ وَتَكَلَّمُوا بِهٰذَا ٱلْكَلَامِ ثُمَّ أُطْبِقَ ٱلتَّابُوتُ وَحُمِلَ إِلَى ٱلْإِسْكَنْدَرِيَّةِ ٭

III.

TRANSLATION INTO ARABIC.

NOTE. The order of the words in the following sentences has been adapted, so far as possible, to that required by the Arabic translation. In addition, however, the student must bear especially in mind the difference of order (§§ 135, 139—142) which marks the cardinal distinction between verbal and nominal sentences (§ 139 note). The square brackets enclose words which in translation should be omitted, while those in curved brackets give the form of the sentence required by the Arabic idiom. — Past and perfect tenses are generally to be rendered by the Arabic perfect, present and future tenses by the Arabic imperf. The extensive use of the (generic) article in Arabic is to be noted. All nouns not in the construct state should have the (definite) article prefixed unless qualified in English by an indefinite article. — So far as lexical the footnotes to the exercises are only *supplementary* to the Glossary. It is, for example, only in special or exceptional cases that "oh" is to be rendered by أَيُّ instead of يَا, and the notes draw attention to such cases. — The apology for violence done to the Queen's English, in the interests of the learner, may be repeated from the first edition, from which the following is in the main reprinted.

A. Nominal Sentences.[1]

1. The glory of the man [is] his sons, and the solicitude of the man [is] his dwelling and his neighbour.—2. The elegance of the man [lies] in his tongue, and the elegance of the woman in her understanding.—3. The liberal [man is] related to God.—4. The worst (of) repentance [is] at the day[2] of resurrection.—5. The love of the world [is] the beginning of every sin.—6. The promise of the king [is] a security.—7. The learned [men are] the heirs of the prophets.—8. Wisdom [is] for the character[3] like medicine for the body.[3]—9. The world [is] the prison of the believer and the paradise of the unbeliever.—10. Contentment [is a part] of[4] the nature[3] of the domestic animals.—11. The malady of covetousness has no (not is[5] for it a) cure; and the disease of ignorance has no (not is for it a) physician.—12. The nutriment of the body[3] [is] (the) beverages and (the) viands, and the nutriment of the under-

---

[1] §§ 139 ff. [2] § 113 a. [3] plur. [4] مِنْ [5] § 50.

standing [is] wisdom and learning.—13. Money has (to money [is]) a difficult entrance and an easy exit.— 14. Verily[1] God [is] forgiving and[2] compassionate.— 15. Verily ye[3] [are] in a manifest error.—16. The nobles of[4] Pharaoh's folk said[5], "Verily this [is] surely[6] a learned enchanter".—17. Verily in that[7] [lies] surely an example for the unbelievers.—18. Flight in its [proper] time [is] better than endurance in its wrong time (in another than its [proper] time).—19. There is no (not[5] [is there]) strength and no (not[8]) power except with[9] God, the High and[10] Mighty [One].— 20. The best of gifts [is] understanding, and the worst of misfortunes [is] ignorance.—

B. The Strong Verb.

21. Jonah went out from the whale's belly.— 22. Zaid killed Muhammed.—23. They gave[15] (beat) Omar a violent beating[14].—24. The direction of prayer was shifted[12] from Jerusalem to Mecca.—25. God knoweth (knowing) what[13] ye are doing.—26. Verily[7] God provides for every one his sufficiency.—27. Learning and money [they] cover up[14] every fault, and poverty

---

[1] § 147 a. [2] § 149. [3] suffix. [4] مِنْ. [5] perf. sing. § 136. [6] § 147 b. [7] § 147 a. [8] § 111. [9] بِ. [10] § 122. [11] § 109. [12] § 136 b. [13] بِمَا, § 56 note a. [14] dual. § 136 d. [15] § 137 b.

and ignorance [they] uncover¹ every fault.—28. They took him away and put him in the bottom of the well.—29. The brothers of Joseph returned² to their father.—30. Why hast thou³ not⁴ washed thy shirt? —31. The most⁵ of mankind are not⁶ grateful².— 32. They⁷ believe not⁸ in⁹ the future life.—33. We made heaven [to be] a [well-]preserved roof.—34. Do not do good out of¹⁰ hypocrisy, and do not leave off [doing] it out of¹⁰ modesty.—

35. Why do ye render waste the cultivated countries?—36. Thereupon we sent Moses and his brother Aaron with our signs to Pharaoh and his nobles; then they declared the two of them¹¹ to be liars.—37. The angels said¹², "O Mary! be obedient to thy Lord and "prostrate thyself; verily¹³ God giveth thee glad "tidings of a word from¹⁵ him; and he¹⁴ [is one] of¹⁵ "those¹⁶ who are placed near [to God], and he shall "talk to mankind in the cradle!"—38. It is not seemly to hurry (not is good the hurrying), except in the marrying of a¹⁷ daughter, and the burying of a¹⁷ dead [man], and the entertaining of a¹⁸ guest.—39. Glorify¹⁹ God in the early morning²⁰ and [late] in the evening²⁰.

---

¹ dual. § 136 d. ² plur. ³ fem. ⁴ لَمْ § 101 c. ⁵ sing. § 127. ⁶ لَيْ. ⁷ pronoun. ⁸ part. ⁹ بِ. ¹⁰ § 113 d. ¹¹ suffix in the dual. ¹² § 136 b. ¹³ § 147 a. ¹⁴ pronoun. ¹⁵ مِنْ. ¹⁶ part. ¹⁷ § 118 c. ¹⁸ § 118 c. ¹⁹ plur. ²⁰ indeterm. accus. § 113 a.

40. Verily the hypocrite has (to the hyp. [belong]) three characteristics; his tongue contradicts his heart, and his speech his action, and his exterior his interior.—41. The men of his people used to sit with him[1] on account of his learning.—42. Verily the holy war [is] incumbent[2] on you.—43. The vehemence of a (the) man[3] [is what] causes him to perish[4].—44. The head of al-Husain the son of Alî was brought into the city[5] of Damascus[6] and was placed before Yazîd.—45. Verily we[7] have become Muslims, so[8] become Muslims ye[9] [also]!—46. Do not talk to one another with disgraceful talk!—47. Every thing has (to every thing [belongs]) an indication; and the indication of understanding [is] reflection, and the indication of reflection [is] being silent.—48. We started off towards Bagdad to bring an action against one another[10] before[11] its[12] governor.—49. The most excellent [kind] of praise [is], "[there is] no[13] god except God!" and the most excellent of [good] works [are] the five[14] prayers; and the most excellent [kind] of character [is] (the) being humble.—50. They fought with one another four days[15], then the Byzantines

---

[1] كَانَ sing., then subject, then the verb in the plur. cf. §§ 89 note c; 136 d. [2] part. [3] غَمْرُ. [4] nominal sent. § 139 d a. [5] § 107. [6] § 128. [7] § 96 d. [8] فَـ. [9] pronoun. [10] part. [11] § 113 b. [12] إِلَى. [13] § 72. [14] § 111. [15] masc. determ. after the noun, § 92 a. [16] § 113 a.

were routed [1].—51. What is disliked in [2] the king [is] the being devoted to (the) pleasures, and the hearing of (the) songs and the spending of (the) time therewith (with that).—52. They said, "O our father! verily we [3] "went away, running races [4], and left Joseph with [5] "our baggage; then the wolf ate him".—53. Observe what [is] in the heart of thy brother by means of his eye, for [6] the eye [is] the title-page of the heart!—54. In the fourth year from the birth of Muhammed the [two] angels [7] cut open [8] his belly and extracted [9] his heart; then they cut it [9] [his heart] open and extracted [9] from it a black clot of blood; thereupon they washed [9] his heart and his belly with snow.—55. They conversed [10] about the case of the Apostle.—

56. Verily God hath (to God [are]) [11] servants whom [12] he distinguishes (he distinguishes them) with his favours.—57. Restrain thyself from meat [13] which [14] causes thee to acquire an indigestion, and [from] an action which [14] occasions thee regret [15].—58. Thou hast fallen in love [16] with a girl, a possessor of beauty [15] and elegance [15].—59. Muhammed said, "Help thy brother, "[whether he be] doing wrong [17] or wronged [17]!" They

---

[1] fem. sing. [2] ل. [3] § 96 d. [4] imperf. merely, § 157 b. [5] عِنْدَ with gen. [6] ف. [7] dual. [8] sing. § 136 a. [9] dual § 136 d. [10] § 137 a. [11] § 147 a. [12] without relative particle § 155. [13] indeterm. [14] without relative particle § 155. [15] indet. [16] § 98 c. [17] § 113 b.

asked, "O Apostle of God! how shall we help him, "[if he be] doing wrong¹?" He said, "By restraining "him from doing wrong!"—60. Do not turn away² a beggar!—61. A man (servant) does not believe, until he love for his neighbour (brother) what³ he loves for himself.—

C. The Weak Verb.

62. A poor [man] begged of me, so I gave him [two] pieces of money⁴.—63. Be mindful of death, for he⁵ takes hold of your forelocks; if⁶ ye fly from him, he overtakes you, and if⁶ ye stay, he seizes you.— 64. Music [is] like the spirit and wine [is] like the body; then through their⁷ coming together is born joy.—65. The Apostle used to⁸ preach to his companions and. to exhort them and to teach them the beauties of character⁹.—66. Verily¹⁰ our [true] friends will¹¹ entrust to us their secrets.—67. The lust¹² of the world entails care and sorrow, and abstinence with regard to it restores the heart and the body.—68. Moses said, "I have brought¹³ you an evidence from your "Lord; so let go¹⁴ along with me the Sons of Israel!" —69. Depend on the Living [one], who does not die!

---

¹ § 113 b. ² contracted § 36. ³ لَـ § 156 and note a. ⁴ dual.
⁵ pronoun with foll. part. ⁶ § 159. ⁷ dual-suffix. ⁸ see p. 61*
note 1. ⁹ pl. determ. ¹⁰ § 147 a. ¹¹ سَـ § 99 a. ¹² ذ. ¹³ § 98 e. ¹⁴ sing.

—70. He pleases me, who makes poetry to [1] show his education, not to [1] make gain, and applies himself to singing to [1] enjoy himself, not to [1] seek for himself [reward]. —71. Demand help of the good (people [2] of the good), and of those that act well (and of the acting well).— 72. Choose [3] whichever of the pages thou wilt!— 73. Supplicate much (make much the supplicating), for thou [4] dost not know when [5] answer [6] will be given thee!—74. Restrain your tongues and lower your glances and guard your continence!—

75. A (the) kingdom is made flourishing through justice and is protected by courage and is ruled through [good] government.—76. [Good] government [is], that [7] the gate of the chief be guarded [8] in the [proper] time of being guarded [9], and opened in the [proper] time of being open [9], and the gatekeeper friendly.—77. Jalâl-al-dîn used not to go to sleep [10] except drunk [11], nor (and not) to arise in the morning except seedy and tipsy [11].—78. It is not seemly for the wise [man], that [12] he address the fool, like as it is not seemly for the sober [man], that he address the drunken [man].—79. People [13] of the world [are] like folk in a ship, who [14] are carried onwards

---

[1] inf. § 113 *d*. [2] § 133. [3] fem. [4] § 96 *d*. [5] مَنِّي. [6] impf. pass. impers. [7] § 148 *b*. [8] كَّى with part. § 110. [9] 61 *c*. [10] see p. 61* note 1. [11] § 113 *b*. [12] § 148 *b*. [13] § 133. [14] §§ 155, 156.

whilst they are sleeping [1].—80. The evil-doer [he] does not consider [2] mankind except [as] evil, because he [3] sees them with [4] the eye of his nature.—81. God elected Abraham [as] an [intimate] friend.[7]—82. Every affair in the world [is] transitory.—83. Wickedness [is] to be feared [5], and no one (not) fears it except the intelligent [man]; and good [is] to be hoped for, and every one [6] seeks it.—84. [To] a man (servant) shall not [8] be given [anything] more ample than endurance.—85. I looked into Paradise, then I saw the most of its inhabitants [to be] the poor; and I looked into hell-fire, then I saw the most of its inhabitants [to be] (the) women.—86. He [9] whose counsel is asked [is] one [10] in whom one confides; and he [10] who asks counsel [is] one [10] who is to be aided.—87. Do not put off [11] the work of to-day till to-morrow [12].—88. Thou dost not [13] find (see) in the creation of God any [14] imperfection.—89. Little which [10] continues [is] better than much which [10] is interrupted.—90. Pharaoh said, "We will [15] kill [16] their sons and spare their women."—91. A Bedouin looked at a gold-piece; then he said, "How small [17] is thy size and how great [17] thy value!"—

[1] § 157 a. [2] § 139 d a. [3] suff. [4] بِ. [5] § 60 c. [6] أَخَذُ.
[7] خَلِيلًا. [8] لَنْ § 100 end. [9] part. [10] part. [11] § 101 b.
[12] indeterm. [13] مَا. [14] مِنْ as used § 141. [15] سَ, § 99 a.
[16] § 19. [17] § 52.

92. The envious [man] is not well-pleased with thee [1], until thou diest!—93. Be [the] tail and be not [the] head! for [2] the tail escapes whilst [3] the head perishes.

D. Various subordinate Sentences.

94. Muhammed said, "Do not anticipate (begin) [4] Jews and Christians by the greeting, but when ye meet one of them [5], (then) [6] force him towards the narrowest place (his narrowest)".—95. When comes to thy knowledge concerning thy brother what is evil, then seek for him excuse; but if thou dost not [7] find [one], then say, "Perhaps he has an excuse."—96. If [8] thou eat little, thou shalt live long.—97. If [8] ye talk in a good manner (make ye good the talk), ye shall enter Paradise.—98. Alî said,—may [9] God be well pleased with him [10]—"O [11] mankind! do not hope except for your Lord, and do not dread [anything] except your transgressions; and be not he ashamed, who [12] doth not know, to [13] learn, and be not he ashamed, who [12] knoweth, to [13] teach!"—99. The subsistence which thou seekest is like the shadow (the likeness of the subsistence... [is] the likeness of the shadow) which moves on along

---

[1] verbal sentence. [2] فَإِنَّ. [3] § 157 nomin. sent. [4] plur. [5] أَحَدَ with gen. § 133 end. [6] § 161 c. [7] §§ 159, 101 c. [8] § 160 b. [9] § 98 d. [10] after the subject. [11] أَيُّهَا § 85. [12] مَنْ. [13] أَنْ with subj.

with thee; thou [1] dost not overtake it in pursuing [2] [it], then when thou turnest [3] away from it, it follows thee! [4]—100. A man said to the Apostle of God: "O Muhammed, give me thy cloak!"; then he threw it down to [5] him; then he said: "I do not [6] want it"; then he [Muh.] said, "May [7] God combat thee! thou didst wish to [8] declare me to be niggardly, but (and) God has not made [9] me [to be] niggardly!"— 101. Whoso [10] longs for Paradise, he is unmindful of lusts [11].—102. That a man [22] give in alms in his lifetime a drachma (the alms-giving [12] of a man—a drachma) [is] better for him than that [13] he give in alms a hundred drachmæ at his death.

103. The Prophet—may God bless [14] him and save him—said, "Whoso [10] drinketh wine in this world, [and] thereupon do not [15] repent, he shall be forbidden it [16] in the future life."—104. If anyone light a lamp in a mosque, then verily [17] the angels [they] will beg forgiveness for him as long as [18] that lamp continues [19] kindled [20].—105. The reed-pen [is] a tree, whose [21] fruit [is] the ideas, and thought [is] a sea, whose [21]

---

[1] pronoun. [2] part. 113 *b*. [3] § 158 *a*. [4] perf. [5] اِلَى. [6] مَا with imperf. [7] § 98 *d*. [8] اَنْ with subj. [9] § 101 *c*.
[10] § 159. [11] determ. [12] inf. [13] مِنْ اَنْ § 148 *b* with subj.
[14] § 11 end. [15] لَمْ §§ 160 *c*, 101 *c*. [16] § 108. [17] § 161 *a*. [18] § 158 *b*.
[19] § 110. [20] part. pass. § 110. [21] § 155. [22] مَرُّ.

pearls [are] wisdom.—106. Verily the dead [man] and he who¹ has no religion (he who no² religion to him) [are] equal³; and there is no² trust in (to) him who¹ has no² piety.—107. Every woman that⁴ has no⁵ modesty [is] like a dish that has no⁵ salt.—108. If anyone's⁶ [whoso, his] tattle is much, his erring is much [also].—109. The anger of the noble [man], although his fire flare up⁷, [is] like smoke of wood⁸ in which [there is] no⁹ blackness.—110. To the ignorant [man] are forgiven¹⁰ seventy¹¹ transgressions, ere to the knowing [man] is forgiven one.

111. Be not¹² like the needle, which¹³ clothes mankind whilst¹⁴ it [is] naked, nor (and) like the wick, which¹⁴ gives light to mankind whilst it is consumed¹⁵.—112. The believer does not escape from the chastisement of God, until he leave off four things, lying, and pride, and niggardliness, and evil thinking (evil of the thinking).—113. It is seemly for the younger [ones] to¹⁶ precede the elders in three places; when¹⁷ they travel by night¹⁸, or wade through a stream, or encounter horsemen.—114. Do not drink (the) poison out of reliance¹⁹ on the antidote which thou hast

---

¹ مَنْ. ² § 111. ³ sing. ⁴ §§ 155, 156. ⁵ part.pass. § 110. ⁶ § 156. ⁷ § 159. ⁸ indeterm. § 155. ⁹ § 111. ¹⁰ § 136 a. ¹¹ § 92 b. ¹² لا with energ. I. § 101 b. ¹³ § 155. ¹⁴ § 157 a. ¹⁵ § 157 a, pron. with imperf. ¹⁶ § 148. ¹⁷ § 158 a. ¹⁸ § 113 a. ¹⁹ § 113 d.

(that which [is] with¹ thee of² the antidote). —
115. Paradise is desirous³ of four [kinds of] folk; the
first⁴ of them⁵ [are] those who have fed⁶ a hungry
[man], and the second [are] those who have clothed⁷
a naked [man], and the third [are] those who fast⁷
in⁸ the month of Ramadân⁹, and the fourth [are]
those who read¹⁰ the Koran.—116. Socrates was asked,
"Why hast thou not¹¹ mentioned in thy law-code the
"punishment of him who kills¹² his brother?" He said,
"I know not that this [is] a thing which exists."—
117. Every thing [it] begins small¹³, thereupon it becomes great, except misfortune¹⁴; for it begins great,
thereupon it becomes small; and every thing [it] becomes
cheap, when¹⁵ it becomes abundant, excepting education;
for¹⁶ when it becomes abundant, it rises in value.

118. After Moses had returned to the Sons of
Israel with the Thora (and along with him [was] the
Thora), they refused to¹⁷ accept it and to do according
to what [was] in it.—119. God commanded Moses to¹⁸
fast thirty¹ˢ days and to purify himself and to purify
his garments, and to come to¹⁹ the mountain, that he
might talk to him and give him the book.—120. After

---

¹ عِنْدَ. ² مِنْ. ³ part. ⁴ masc. ⁵ suffix in fem. sing.
⁶ perf. sing. ⁷ imperf. sing. ⁸ § 113 *a*. ⁹ § 128. ¹⁰ imperf.
sing. ¹¹ § 101 *c*. ¹² § 159. ¹³ § 113 *b*. ¹⁴ accus. § 151. ¹⁵ § 158 *a*.
¹⁶ with suff. § 96 *d*. ¹⁷ أَنْ with subj. ¹⁸ § 113 *a*. ¹⁹ إِلَى.

Damascus was taken [1], much folk [2] of [3] its inhabitants joined Heraclius, whilst [4] he was in [5] Antioch.—121. A certain one of the wise men said, Nothing (not) repels the onslaught of the conquering enemy like [6] being submissive and giving way, like as [7] green plants are safe from the vehement wind through their pliancy, because they [8] turn along with it, as (how) [9] it turns.—122. They disagree [10] concerning Waraḳa; and of [11] them [there are] those who assert [15] that [12] he died a Christian [13] and did not [14] reach the appearance of the Prophet; and of [11] them [there are] those who are of opinion [15] that [12] he died a Muslim.—123. O [ye two] companions of the prison! as to the one of you [16], he shall serve to his lord wine [17], and as to the other, he shall be crucified, then shall [18] the birds eat of [12] his head; the affair is decreed [19] concerning which ye inquire!—124. The Apostle wrote to chieftains [17] of [11] the tribes, inviting [20] them to become Muslims [21].—125. A wise [man] was asked, "What [is] the thing, which [it] is not good that it be said, although it be [22] right?" He said, "A man's eulogizing himself [23]".—126. Woe to

---

[1] fem. § 136 b. [2] بَشَرٌ coll. [3] مِنْ. [4] § 157 a. [5] بِ. [6] مِثْلُ as subject, § 145 b. [7] كَمَا أَنَّ § 147 a. [8] sing. suff. [9] § 159. [10] § 98 b with قَدْ, § 137 a. [11] مِنْ. [12] § 147 a. [13] § 113 b. [14] § 101 c. [15] § 98 b. [16] أَحَدُ w. dual suff. § 133. [17] indeterm. [18] fem. sing. § 136 c, 2. [19] § 98 b. [20] § 99 b. [21] infin. determ. [22] كَانَ § 159. [23] § 131 w. Acc.

[him] who converses with lying, that he may make the people laugh by it!—127. This (the) world and the future life [are] as the East and the West; when thou approachest one of them¹, thou dost recede from the other.—128. Fear ye God in secret² and do not enter into what is not lawful for you!—129. The devotee without learning [is] like the ass of the mill³, who⁴ goes around and does not⁵ get through (cut) the distance.—130. The eye of hate [it] draws forth every fault, and the eye of love [it] does not find the faults.

E. Anecdotes.

131. An astrologer was being crucified; then he was asked⁶, "Hast thou⁷ seen this in thy star?" Then he said, "I saw a raising up⁸, however I did not⁹ know that it [was to be] upon a piece of wood."

132. A man knocked at the door of¹⁰ 'Amr the son of 'Ubaid; so he said "Who [is] this?" He said, "I." He ['Amr] said, "I do not know (I am not I know¹¹) among our friends (brothers)¹² [any] one¹³, whose name [is] I."

133. (The) thieves came¹⁴ in upon Abû Bekr al-Rabbânî, seeking¹⁵ something (a thing), and he saw

---

¹ dual suffix. ² determ. ³ § 123, note. ⁴ § 155 note. ⁵ § 157*b* لا w. impf. ⁶ 137 *a*. ⁷ with interrog. part. هَلْ. ⁸ 73 *c* end. ⁹ § 101 *c*. ¹⁰ عَلَى. ¹¹ لَيْسَ § 50 and impf. ¹² order § 131 *b*. ¹³ أَحَدٌ. ¹⁴ § 136 *a*. ¹⁵ § 157 *b* imperf. alone.

them going around¹ in the house. Then he said, "O young men! This which ye are seeking² in the night³ we have⁴ already sought² in the day-time, but have not⁵ found it!" So they laughed and went out.

134. It is related⁶, that⁷ a certain one of the polite scholars eulogized a certain one of the princes; so he commanded [that] to him an [ass's] saddle and saddle-girth [should be given]. So he took them⁸ on⁹ his shoulder and went out from his presence¹⁰. Then a certain one of his companions saw him, then said, "What [is] this?" He said, "I eulogized the prince with the most beautiful of my poems, then he invested me with [something] of¹¹ the most glorious of his dresses".

135. Al-Muġîra, the son of Šuʿba said: No one (not)¹² has deceived me except (another than) a youth of¹³ the sons of al-Ḥâriṯ. For I mentioned a woman of theirs (of¹¹ them), that¹³ I should marry her; then he said, "O¹⁴ Prince! [There is] no good¹⁵ for thee in her." So I said, "And why [not]?". He said, "I saw a man kissing¹⁶ her." So I turned from her; then the young man married her. So I reproached him and said, "Didst thou not¹⁷ inform me that thou¹⁸ hadst

---

⁷ imperf. ² with suffix. ³ § 118 a. ⁴ § 98 e. ⁵ مَا § 150 a. ⁶ § 98 b. ⁷ أَنَّ. ⁸ dual suffix. ⁹ عَلَى. ¹⁰ مِنْ عِنْدِهِ. ¹¹ مِنْ. ¹² 101 c. ¹³ لِ. ¹⁴ أَيُّهَا. ¹⁵ § 111. ¹⁶ imperf. ¹⁷ أَلَمْ § 101 c. ¹⁸ أَنَّ with suff.

seen a man kissing her?" He said, "Yes, I saw her father kissing her."

136. Al-Ḍaḥḥâk the son of Muzâḥim said to a Christian, "[How would it be] if¹ thou wert to become a Muslim?" He said, "I have not² ceased loving³ Islâm⁴, except that⁵ my love for wine⁶ prevents me from it." So he said, "Become a Muslim and drink it!" So after he had become a Muslim, he said to him, "Thou hast⁷ become a Muslim, so if thou drink it⁸, we shall chastise thee; and if thou apostatize, we shall have thee killed⁹, so choose for thyself". Then he chose Islâm and his Islâm was good. So he had taken¹⁰ him by stratagem.

137. A Bedouin stole a purse in which (it) [were] pieces of money¹¹, thereupon he entered the mosque to pray¹²; and his name was¹³ Moses. Then the leader of prayer recited, "And what is that¹⁴ in¹⁵ thy right hand, Oh Moses¹⁶?" So he said, "By God, verily thou [art] an enchanter!" Thereupon he threw away the purse and went out.

138. A man claimed the (a) gift of prophecy in the days¹⁷ of al-Rašid. So after he had appeared

---

¹ لَوْ § 102. ² مَا with perf. ³ § 110 with indeterm. part. ⁴ § 132 end. ⁵ §§ 147 c, 148 أَلَّا أَنَّهُ with foll. verbal sentence. ⁶ § 131. ⁷ § 98c with قَدْ. ⁸ § 159. ⁹ § 17, note b. ¹⁰ perf. ¹¹ indeterm. ¹² § 99 b. ¹³ كَانَ. ¹⁴ fem. ¹⁵ بِ. ¹⁶ Surah 20, 18. ¹⁷ § 113 a.

before him [the Caliph], he asked him, "What [is that] which is said of thee?" He said, "that I[1] am a noble prophet." He asked, "But what[2] indicates the truth of thy claim?" He said, "Demand what[3] thou wilt"[4]. He said, "I wish that[5] thou make these[6] beardless slaves, [who are] standing[7] [there] this moment[8] [to be furnished] with beards[9]." Then he looked down for a while[10], thereupon he raised his head and said, "How is it lawful that I make these[11] beardless [ones to be furnished] with beards[9] and alter these[6] beautiful[12] forms? but[13] I will make the bearded ones (owners of beards) beardless in one twinkling." So al-Rašîd laughed at him and pardoned him and commanded a present [to be given] to him.

139. A person pretended to prophecy[14]; then they besought of him in[15] the presence of al-Ma'mûn a miracle. So he said, "I will cast for you a pebble into the water, then it will dissolve". He [al-Ma'mûm] said, "We are[16] content." So he brought out a pebble [which he had] along with him[17], then cast it into the water; then it dissolved. So they said, "This[18] is a

---

[1] § 96 d. [2] أَيُّ شَيْءٍ. [3] § 5, note b. [4] perf. § 159. [5] أَنْ.
[6] § 120 d; the dem. in sing., the adj. in broken pl. [7] determ. § 120 a. [8] § 118 a. [9] indeterm. [10] § 113 a. [11] plur. [12] § 120 fem. sing. [13] وَأَنَّمَا. [14] § 22. [15] بِ [16] § 98 c. [17] مَعَهُ § 121 a.
[18] § 143.

trick; however, we will give¹ thee a pebble of our own², and let³ it dissolve!" Then he said, "Ye are not⁴ more illustrious⁵ than Pharao and I am not (and not I⁶) mightier in wisdom⁷ than Moses, and Pharao did not⁸ say to Moses, 'I am not⁹ content with what thou doest¹⁰ with thy staff, so that¹¹ I will give thee a staff of my own¹², which¹³ thou shalt make [into] a serpent.'" So al-Ma'mûn laughed and let him pass on.

140. It is said¹⁴ that Abû Dulâma¹⁵ the poet was standing¹⁶ before al-Saffâḥ on¹⁷ a certain day (a certain one of the days). Then he said to him, "Ask of me what thou dost want (thy want)!" So Abû Dulâma said to him, "I want a hunting-dog". So he said, "Give ye it¹⁸ to him!" Then he said, "And I want a horse, on¹⁹ which I may go forth to hunt." He said, "Give ye it to him!" He said, "And a page²⁰, who²¹ will lead the dog and hunt with him." He said, "And give ye him a page!" He said, "And a slave-girl²², who²³ will prepare the game and give us to eat of it." He said, "Give ye him a slave-girl!" He said, "These,

---

¹ imperf. ² مِنْ عِنْدِنَا. ³ imper. of. ودع w. suff.; then impf. لَمْ ⁴ § 110. ⁵ § 63 b. ⁶ وَلَا أَنَا. ⁷ § 113 c. ⁸ § 101 c. ⁹ لَمْ.
¹⁰ § 156. ¹¹ حَتَّى with subj. ¹² مِنْ عِنْدِي. ¹³ §§ 155—56.
¹⁴ § 98 c. ¹⁵ 147 a. ¹⁶ كَانَ with part. § 110. ¹⁷ فِي. ¹⁸ with إِيَّا, which stands last, § 54 b. ¹⁹ عَلَى (after the verb) § 155.
²⁰ accus. ²¹ § 155. ²² accus. ²³ § 155.

O Prince of the Believers! have need of ([there is] no¹ escape for them from) a dwelling, which² they may inhabit." So he said, "Give ye him a dwelling, which² will contain them!" He said, "And if they have not (and if not is³ to them) an estate, then wherefrom shall they live?" He said, "I grant⁴ thee ten cultivated⁵ estates and ten waste estates⁵." He said, "And what [are] the waste⁵ [ones] O Prince of the Believers?" He said, "In which⁶ [there are] no plants⁷." He said, "I⁴ grant thee, O Prince of the Believers, a hundred⁸ waste estates of⁹ the deserts of the Sons of Asad." Then he laughed at him and said, "Make them¹⁰ all of them¹⁰ cultivated!¹⁰"

141. It is related¹¹, that Harûn al-Rašîd had (that to H. was¹²) a black slave-girl, of ugly mien¹³. Now he scattered one day gold-pieces¹⁴ among (between) the slave-girls; so the slave-girls set about¹⁵ gathering¹⁶ up the gold-pieces, whilst¹⁷ that slave-girl stood still, looking¹⁸ at the face of al-Rašîd. Some one asked (it was asked), "Dost thou¹⁹ not pick up the

---

¹ § 111. ² §§ 155—56. ³ لَمْ تَكُنْ. ⁴ § 98 c with قَدْ. ⁵ § 87 a. ⁶ مَا and prep. with pronoun at the end of the sentence. ⁷ § 111. ⁸ § 92 c. ⁹ مِنْ. ¹⁰ fem. sing. ¹¹ § 98 c. ¹² أَنَّهُ كَانَ لِ § 147 c. ¹³ determ. § 134. ¹⁴ indeterm. ¹⁵ § 136 a. ¹⁶ §§ 152 note b, 136 d (impf. pl. fem.) ¹⁷ § 157 a with part. ¹⁸ § 157 b impf. alone. ¹⁹ أَلَا w. impf. fem.

gold-pieces?" Then she said, "Verily what[1] they seek [is] the gold-pieces, but (and) what[1] I seek [is] the owner of the gold-pieces." Then her speech pleased him; so he placed her near [to him] and brought good upon her. Then the report got to the grandees, that[2] Harûn al-Rašîd was enamoured[3] of a black slave-girl. So after that had come to his knowledge, he sent for the whole of the grandees, until he had assembled[4] them in his presence[5]. Then after he had commanded the bringing in[6] of the slave-girls, he gave every one of[7] them a goblet of[7] chrysolite[8] and commanded it to be thrown down[6]. But they declined [doing it] in a body (as a whole[9]). Then the turn came to (the affair got to) the ugly slave-girl; but she threw down the goblet and broke it. So they said, "Look[10] at this girl, her name [is] ugly, and her manner [is] ugly, and her action [is] ugly". Then said to her the Caliph, "Why then didst thou break[11] it"? Then she said, "Thou didst[12] command me to break it[13]; so I was of opinion that[14] in[15] its being broken [lay] a detriment[16] with regard to the

---

[1] part. pass. with suffix. [2] بِأَنَّ. [3] imperf. [4] § 152, note c. [5] عِنْدَهُ. [6] بِ with infinitive § 131. [7] مِنْ § 119 a. [8] determ. [9] § 113 b. [10] plur. [11] 2nd. pers. fem. perf. w. suff. § 53 a. [12] § 98 e. [13] بِ with inf. [14] أَنَّ. [15] فِي. [16] § 147 a.

treasure of the Caliph, and in its not being broken (in the lack of its being broken) a detriment [1] with regard to his command; and the detriment with regard to the first is fitter to keep intact [2] the inviolability of the command of the Caliph. And I was of opinion that in its being broken [lay] my being called (qualified [3] as [4]) the crazy [one], and in keeping it intact my called being (qualified [3] as [4]) the disobedient [one]; and the first [is] more agreeable to me than the second." Then the grandees found [5] that [6] to be beautiful of [7] her and praised her for [8] it and excused the Caliph for [9] loving her. And God knows best ([is] most knowing [10]).

---

[1] § 147 a. [2] § 113 d, indeterm. inf. with following لِ § 131. [3] § 61 c. [4] بِ. [5] § 136 a. [6] at the end. [7] مِنْ. [8] عَلَى. [9] فِي. [10] elative.

GLOSSARY A.

pl. = plural, see §§ 88—90. The numbers within parentheses after the broken plurals refer to the forms as numbered in these sections.

Aaron هَارُون.
Abraham إِبْرَهِيم.
abstinence زُهْد.
Abū Bekr al-Rabbāni أَبُو بَكْرِ ٱلرَّبَّانِيّ.
Abū Dulāma أَبُو دُلَامَة.
abundant *see* much.
accept (to) قَبِلَ *impf. a.*
acquire (to cause to) كسب *IV with two accus.*
act well (to) حسن *IV.*
action فِعْل. *See also* bring.
address (to) خطب *III.*
affair أَمْر.
after, after that *conj.* لَمَّا § 98 *f.*

after *prep.* خَلْفَ.
agreeable to *elat.* أَحَبّ
  with إِلَى.
aid (to) عان *med.* و *IV*
  with acc.
'Alī عَلِيّ.
all كُلّ *with determ. noun*
  or suffix § 119 *b.*
alms (to give in) صدق
  V with بِ *of the gift.*
along with *prep.* مَعَ.
already قَدْ § 98 *e.*
alter (to) غار *med.* ى *II.*
although وَإِنْ § 159.
among فِي.
ample وَاسِع *elat.* § 63 *b.*

'Amr عَمْرو § 90 n.
and وَ.
angel مَلَأَك pl. مَفَاعِلَة (28).
anger غَضَب.
animal (domestic) بَهِيمَة pl. فَعَائِل (25).
another than غَيْر with following gen.
answer (to give) to جاب X med. و with لِ.
antidote دِرْيَاق.
Antioch أَنْطَاكِيَة.
any مِنْ (prep.), cf. § 141.
apostatize رَدّ VIII.
apostle رَسُول.
appear (to) مَثَل.
appearance ظُهُور.
apply oneself to (to) عطا VI with acc.
approach (to) قَرُبَ impf. u, with مِنْ.

arise (to, in the morning) صبح IV.
as see like.
as to أَمَّا with nom. and ف in the apodosis.
Asad أَسَد.
ashamed (to be) حىِ X § 49 c.
ask (to) قَالَ med. و with لِ. — to ask something of سَأَل impf. a, with two acc. § 38 b.
ass حِمَار.
assemble (to) جَمَع impf. a.
assert (to) زَعَم impf. u.
astrologer مُنَجِّم.
at (one's house) prep. عِنْدَ.
Bagdad بَغْدَاذ.
baggage مَتَاع.
be, exist (to) كَانَ med. و. — not to be لَيْسَ § 50.
beard لِحْيَة pl. فِعَل (3); cf. § 71 b.

beardless أَمْرَدُ pl. فُعْلُ (1).
beat (to) ضَرَبَ impf. i, inf. ضَرْبٌ.
beauty حُسْنٌ. — beauties مَحَاسِنُ.
beautiful حَسَنٌ fem. ةَ; elat. § 63 b. — to find to be beautiful حسن X.
because لِأَنَّ § 147 a.
Bedouin أَعْرَابِيٌّ.
before (of place) = between the two hands of (dual stat. constr.).
beg of (to) سَأَلَ impf. a, with acc.
beggar part. act. of سَأَلَ.
begin, begin with (to) بَدَأَ impf. a, with acc.
beginning رَأْسٌ (lit. head).
believe (to) امن IV; — believer id. part. act.
belly بَطْنٌ.

beseech of (to) طلب III with acc. of person and ب of thing.
best elat. of good.
better elat. of good.
between بَيْنَ.
beverage مَشْرَبٌ pl. مَفَاعِلُ (23).
birds coll. طَيْرٌ.
birth مَوْلِدٌ.
black أَسْوَدُ fem. § 74 b.
blackness سَوَادٌ.
bless (to) صلا II with على.
body جَسَدٌ pl. أَفْعَالٌ (17). بَدَنٌ (no. 67).
book كِتَابٌ.
born (to be) ولد V.
bottom غَيَابَةٌ.
break (to) كَسَرَ impf. i.
bring (to) جَاءَ ب med. ى.
— to bring an action against one another

حكم VI. — to bring in
حضر IV. — to bring
into دخل IV. — to
bring out خرج IV. —
to bring upon أتى IV
with عَلى.

brother أَخٌ § 90 a, c; pl.
§ 88, 5; pl. when =
"friends" § 88, 21.

bury (to) دَفَنَ impf. i, inf.
دَفْنٌ.

but فَ.

by, by means of بِ; in
oaths = وَ w. the gen.
§ 95 i.

Byzantines (the) coll. اَلرُّومُ.

Caliph خَليفَةٌ.

care هَمٌّ.

carry onwards (to) سَارَ
med. ى, with بِ.

case خَبَرٌ.

cast (to) طَرَحَ impf. a.

cease (to) زَالَ med. و, (for
زَوَلَ § 42 d, § 44).

certain one (a) بَعْضٌ with
pl. of follow. noun.

character خُلْقٌ pl. أَفْعَالٌ
(17).

characteristic عَلَامَةٌ.

chastise (to) حَدَّ impf. u.

chastisement عَذَابٌ.

cheap (to become) رَخُصَ
impf. u.

chief رَئِيسٌ pl. فَعَلَاءُ (20).

choose (to) خَارَ med. ى
VIII.

chrysolite يَاقُوتٌ.

Christian نَصْرَانِيٌّ pl. فَعَالَى
نَصَارَى (29).

claim (to) دَعَا VIII § 25,
note.

claim دَعْوَى.

cloak رِدَاءٌ.

clot of blood عَلَقَةٌ.

clothe (to) كَسَا impf. u.
city مَدِينَة.
cognizant of عَلِيم بِ.
combat (to) قتل III.
come (to) أَتَى impf. i. — to come to one's knowledge (concerning) بَلَغَ impf. u, with acc. (and عَنْ). — to come in upon دَخَلَ impf. u, with عَلَى. — to come out from خَرَجَ impf. u, with مِن.
come together جمع VIII.
command (to) أَمَرَ impf. u. — to command anyone to do a thing, id. with acc. and أَنْ with the subj. — to command anything to be given to anyone, id. with لِ of pers. and بِ of thing. — to command any thing to be done, id. with بِ and infin.
command أَمْر.
companion صَاحِب pl. أَفْعَال (17).
compassionate رَحِيم.
concerning فِي.
confide in (to) أمن VIII.
conquering part. act. of قَهَرَ.
consider as (to) ظَنَّ impf. u, with acc.
consumed (to be) حرق VIII.
contain (to) جَمَعَ impf. a.
content (to be) رَضِيَ impf. a. — to be content with, id. with بِ.
contentment تَنَاعَة.
continence فُرُوج (pl. of فَرْج).

F*

continue (to) دَامَ med. و
§ 110.
contradict (to) خلف III.
converse (to) حدث V. —
to converse about, id.
with بِ.
counsel (to ask) شاور med.
و X.
country بَلَدٌ pl. فِعَالٌ (9).
courage شَجَاعَةٌ.
cover up (to) سَتَرَ impf. u.
covetousness حِرْصٌ.
cradle مَهْدٌ.
crazy part pass. of جَنَّ
fem. ة.
creation خَلْقٌ.
crucify (to) صَلَبَ impf. i.
cultivated part. act. of
عمر fem. ة.
cure شِفَآءٌ.
cut (to) قَطَعَ impf. a. —
to cut open شَقَّ impf. u.

ad-Ḍaḥḥāk اَلضَّحَّاكُ.
Damascus دِمَشْقُ.
daughter بِنْتٌ § 90 i.
day يَوْمٌ pl. أَيَّامٌ §§ 88, 17;
90 s. — one day يَوْمًا.
to-day اَلْيَوْمَ.
day-time نَهَارٌ.
dead مَيِّتٌ.
death مَوْتٌ.
deceive (to) خَدَعَ impf. a.
decline (to) منع VIII.
decree (to) قضى impf. i.
demand (to) a thing سَأَلَ
impf. a, with عن § 38 b.
depend on (to) وكل V, with
عَلَى.
desert فَيْفَآءٌ pl. فَعَالٍ (26);
فَيَافٍ.
desirous of (to be) شَاقٍ
med. و VIII, with إِلَى or
عَلَى.
detriment نَقْصٌ.

devoted to (to be) همك VIII with فِي.

devotee *part. act. of* عبد V.

die (to) مَاتَ *med.* و.

difficult عَسِيرٌ.

disagree (to) خلف VIII.

disease دَاءٌ.

disgraceful قَبِيحٌ.

dish طَعَامٌ.

dislike (to) كَرِهَ *impf. a.*

disobedient *part. act. of* عصى.

dissolve (to) ذَابَ *med.* و.

distance مَسَافَةٌ.

distinguish (to) خَصَّ *impf. u.*

do (to) عَمِلَ *impf. a;* فَعَلَ *impf. a* (no. 139). — to do according to عَمِلَ *with* بِ.

dog كَلْبٌ; hunting-dog كَلْبُ صَيْدٍ.

domestic *see* animal.

door بَابٌ.

drachma دِرْهَمٌ.

draw forth (to) برز IV.

dread (to) خَشِيَ *impf. a.*

dress مَلْبَسٌ *pl.* فَعَالِلُ (23).

drink (to) شَرِبَ *impf. a.*

drunk, drunken سَكْرَانُ.

dwelling دَارٌ (*fem.*).

early *see* morning.

East مَشْرِقٌ.

easy يَسِيرٌ.

eat (to) أَكَلَ *impf. u; imp.* § 38 b. — to give to eat of طعم IV *with acc. pers. and* مِنْ.

education أَدَبٌ. — to show one's education ادب V.

elder أَكْبَرُ *pl.* أَفَاعِلُ (23).

elect (to) صفى *VIII.*
elegance جَمَال.
enamoured of (to be) عَشِقَ *impf. a, with acc.*
enchanter سَاحِر.
encounter (to) لقى *III.*
endurance صَبْر.
enemy عَدُوّ.
enjoy oneself (to) طرب *V.*
entail (to) ورث *IV.*
enter (to) دَخَلَ *impf. u.* see § 107 *note.*
entertain (to) قَرَى *impf. i. inf.* قِرَآء.
entrance مَدْخَل.
entrust (to) anyone with ودع *X with two accus.*
envious *part. act. of* حسد.
equal سَوَآء.
ere, *conj.* قَبْلَ أَنْ § 100.
err (to) غَلِطَ *inf.* غَلَط.
error ضَلَال.

escape (to) نَجَا *impf. u.*
escape بُدّ.
estate ضَيْعَة *pl.* ضِعَال (9).
eulogize (to) مَدَحَ *impf. a.; id. VIII* (no. 134).
evening (late) عَشِيّ.
every كُلّ *with indeterm. noun.* § 119 *b.*
evidence بَيِّنَة.
evil (to be) سَآءَ *med.* و.
— to do evil id. *IV.* —
evil-doer *part act. of id. IV.*
evil سُوء.
example عِبْرَة.
excellent فَاضِل *elat.* § 63 *b.*
except إِلَّا (= إِنْ لَا) § 151.
— except that إِلَّا أَنَّهُ § 147 *c.*
excepting مَا خَلَا *with acc.*
excuse (to) عَذَرَ *impf. i.*
excuse عُذْر.

GLOSSARY A.

exhort (to) وَعَظَ impf. i, § 40 a.
exist (to) كَانَ med. و.
exit خَرَجَ.
exterior عَلَانِيَةٌ.
extract (to) خرج X.
eye عَيْنٌ fem. § 72.
face وَجْهٌ.
fast (to) صَامَ med. و.
father أَبٌ § 90 a.
fault عَيْبٌ pl. عُيُوبٌ (10).
favour نِعْمَةٌ pl. فِعَلٌ (3).
fear (to) خَافَ med. و, impf. a, § 42 d.
feed (to) طعم IV.
fight (to) with one another قتل VI.
find (to) وَجَدَ impf. i, § 40 a.
fire نَارٌ.
first أَوَّلُ.
fit وَلِيٌّ elat. أَوْلَى.

five خَمْسٌ §§ 91, 92 a.
flare up (to) أَجَّ I.
flight هَرَبٌ.
flourishing (to make) عَمَرَ impf. u.
fly (to) from فَرَّ impf. i, with مِنْ.
folk قَوْمٌ pl. § 88, 17; بَشَرٌ coll. (no. 120).
follow (to) تَبِعَ impf. a.
fool part act of جهل.
for prep. لِ § 95 h; conj. فَإِنَّ § 96 d.
forbid (to) a thing to anyone حَرَمَ impf. i, with two accus.
force (to) ضر VIII. § 25 note.
forelock نَاصِيَةٌ pl. فَوَاعِلُ (24).
forgive (to) غَفَرَ impf. i.
forgiving غَفُورٌ.

forgiveness (to beg) غفر X.
form صُورَة pl. فُعَل (4).
four أَرْبَع §§ 91, 92 a.
fourth رَابِع.
friend صَاحِب (see p. 85*).
— of God = Abraham خَلِيل.
— intimate صَدِيق pl. أَفْعِلَاء (18).
friendly لَطِيف.
from prep. مِنْ.
fruit ثَمَرَة.
future life see life.
gain (to make) كسب V.
game صَيْد.
garment ثَوْب pl. فِعَال (9).
gate بَاب.
gate-keeper بَوَّاب.
gather up (to) لقط VIII.
get to (to) نهى VIII with إِلَى. — to get through قطع impf. a.

gift مَوْهِب pl. مَفَاعِل (23).
see also prophecy.
girl جَارِيَة.
give (to) عطا IV with two acc. — to give way inf. خُضُوع.
glad see tidings.
glance بَصَر pl. أَفْعَال (17).
glorify (to) سبح II.
glorious فَاخِر elat. § 63 b.
glory شَرَف.
go round (to) دَارَ med. و.
— to go away ذَهَبَ impf. a. — to go on مَشَى impf. i. — to go out خَرَجَ impf. u. — to let go رسل IV.
goblet قَدَح.
god إِلَه; God اَللّٰه, by God وَاَللّٰهِ.
gold-piece دِينَار pl. § 90 k.
good noun and adj. خَيْر.

GLOSSARY A.

elat. id. — to be good حَسُنَ impf. u. — to make good طاب med. ى IV.

government رِئَاسَةٌ or رِيَاسَةٌ.

governor وَلِيٌّ.

grandee مَلِكٌ pl. § 88, 10.

grant (to) قطع IV with two accus.

grateful (to be) شَكَرَ impf. u.

great كَبِيرٌ. — to be, become great كَبُرَ impf. u.

green (fresh) رَطْبٌ.

greeting سَلَامٌ.

guard (to) صَانَ med. و; inf. صَوْنٌ.

guest ضَيْفٌ.

hand يَدٌ § 90 r.

al-Ḥārit ٱلْحَارِثُ.

Hārūn ar-Rashīd هَارُونُ ٱلرَّشِيدُ.

hate بُغْضٌ.

have (to), *is expressed by the subject in the dative (with* لِ*) followed by the object in the nom. (as* لَهُ مَالٌ *he has money); occasionally a form of* كَانَ *to be stands before the subject (as* كَانَ لَهُ مَالٌ *he had money).* — not to have *either as in the last example, but with* لَيْسَ *(§ 50) instead of* (لَيْسَ لَهُ مَالٌ) كَانَ *or* لَا *with following object (§ 111) and dative of subject* (لَا مَالَ لَهُ).

he هُوَ § 12 a. — he who مَنْ § 14 b.

head رَأْسٌ.

hear (to) سَمِعَ impf. a, inf. سَمَاعٌ.

heart قَلْبٌ pl. فُعُولٌ (10).

heaven سَمَآءٌ pl. سَمَوَاتٌ § 76 b.

heir part. act. of ورث pl. وَرَثَةٌ (6).

hell-fire اَلنَّارُ.

help (to) نَصَرَ impf. u. — to demand help of med. و, X with بِ.

Heraclius هِرَقْلُ.

high عَلِيٌّ.

holy see war.

hope for (to) رَجَا impf. u, with acc.

horse دَابَّةٌ.

horsemen coll. خَيْلٌ.

house بَيْتٌ.

how كَيْفَ.

however وَلَاكِنْ with follg. verb.

humble (to be) وضع VI.

hungry part. act. of جَاعَ med. و.

hundred مِائَةٌ §§ 91, 92 c.

hunt (to) صَادَ med. ى. — to go forth to hunt id. V.

hunt, chase صَيْدٌ.

hurry (to) عجل II.

al-Husain اَلْحُسَيْنُ.

hypocrisy (religious) رِيَآءٌ.

hypocrite part. act. of نفق III.

I أَنَا.

ignorance جَهْلٌ.

ignorant part. act. of جهل.

idea مَعْنًى pl. مَفَاعِلُ (23).

if إِنْ § 159; in hypothetical clauses لَوْ with the perf. — if anyone مَنْ § 159.

illustrious جَلِيلٌ elat. § 63 b.

imperfection inf. of فَاتَ med. و, VI.

in prep. فِي.

incumbent on (to be) وَجَبَ impf. i, with عَلى § 40 a.

indicate (to) دَلَّ *impf. u*, with عَلَى.

indication دَلِيلٌ.

indigestion بَشَمٌ.

inform (to) خبر *IV.*

inhabit (to) سَكَنَ *impf. u.*

inhabitants أَهْلٌ.

inquire concerning (to) فتى *X with* فِي.

intelligent *part. act. of* عقل.

interior سَرِيرَةٌ.

interrupted (to be) قطع *VII.*

intimate *see* friend.

into *prep.* فِي.

invest (to) anyone with خَلَعَ *impf. a*, with عَلَى *of pers. and acc. of thing.*

inviolability حُرْمَةٌ.

invite to (to) دَعَا *impf. u*, with إِلَى.

Islam اَلْإِسْلَامُ.

Israel إِسْرَائِيلُ.

Jalāl ad-dīn جَلَالُ ٱلدِّينِ.

Jerusalem اَلْقُدْسُ.

Jews (the) *coll.* اَلْيَهُودُ.

join (to) لَحِقَ *impf. a*, with بِ.

Jonah يُونُسُ.

Joseph يُوسُفُ.

joy سُرُورٌ.

justice عَدْلٌ.

keep from (to) مَنَعَ *impf. a*, with *acc. and* مِنْ.

keep intact (to) بقى *IV.*

kill (to) قَتَلَ *impf. u.*

kindle (to) وَقَدَ *impf. i.*

king مَلِكٌ.

kingdom مَمْلَكَةٌ.

kiss (to) قبل *II.*

knock (to) at the door of

دَقّ *impf. u, with* عَلَى *of pers. and acc. of door.*

know (to) عَلِمَ *impf. a;* عَرَفَ *imp. i* (no. 132), دَرَى *impf. i* (no. 73).

knowing *part. act. of* علم; *elat.* § 63 *b.*

Koran اَلْقُرْآن.

lack عَدَم.

lamp سِرَاج.

laugh (to) ضَحِكَ *impf. a.* — to laugh at id. *with* مِنْ — to make laugh id. *IV with* بِ *of means.*

law-code شَرِيعَة.

lawful (to be) حَلَّ *impf. i.*

lead (to) قَادَ *med.* و.

leader *see* prayer.

learn (to) علم *V.*

learned عَلِيم *pl.* فُعَلَاء (20).

learning عِلْم.

leave, leave off (to) تَرَكَ *impf. u.*

let (to) وَدَعَ *impf. a* § 40 *a.*

liar (to declare anyone to be a) كذب *II.*

liberal سَخِيّ.

lie, tell a lie (to) كَذَبَ *impf. i; inf.* كِذْب.

life (the future, next world) اَلْآخِرَة.

life-time حَيَاة.

light (to) سرج *IV.* — to give light to ضاء *med.* و, *IV, with* لِ.

like (like as) *prep.* كَ; *conj.* كَمَا أَنْ (*with vb.sent.*), كَمَا (*nom. sent.*).

likeness مَثَل.

little قَلِيل.

live (to) عَاشَ *med.* ى.

living حَيّ.

long طَوِيل.

long for (to) شانَ *med.* و
VIII, *with* عَلَى.

look at (to) نَظَرَ *impf. u,
with* إِلَى.—to look down
طرق *IV.*— to look into
طلع *VIII, with* فى § 25,
note.

lord رَبّ.

love, fall in love with (to)
حبّ *IV, with acc.*

love حُبّ.

loving *inf.* مَحَبَّة.

lower (to) غَضَّ *impf. u.*

lust رَغْبَة.— lusts شَهَوَات.

make, make to be (to) جَعَلَ
impf. a, (with two accus.).
— to make (poetry) قَالَ
med. و.

malady سَقَام.

al-Ma'mūn اَلْمَأْمُون.

man رَجُل *pl.* رِجَال (9); مَرْء

antith. to woman (nos.
2, 43, 102), § 90 *e.*

manifest *part. act.* بَان
med. ى *IV.*

mankind *coll.* اَلنَّاس.

manner وَضْع.

marry (to) زَاج *med.* و, *V.*

Mary مَرْيَم.

meat لَحْم.

Mecca مَكَّة.

medicine طِبّ.

meet (to) لَقِىَ *impf. a.*

mention (to) ذَكَرَ *impf. u.*

mien مَنْظَر.

mighty عَظِيم *elat.* § 63 *b.*

mill طَاحُونَة.

mindful of (to be) ذَكَرَ
impf. u, with acc.

miracle مُعْجِزَة.

misfortune مُصِيبَة *pl.*
نَعَائِل (25).

modesty حَيَآءٌ.
moment (this) اَلسَّاعَةَ.
money مَالٌ. — piece of money دِرْهَمٌ pl. تَعَالِلُ (23).
month شَهْرٌ.
morning (early) بُكْرَةٌ.
morrow, to-morrow غَدٌ.
Moses مُوسَى.
mosque مَسْجِدٌ.
most elat. of much.
mountain جَبَلٌ.
much كَثِيرٌ elat. § 63 b. — to be much, abundant كَثُرَ impf. u. — to make much كثر IV.
al-Mugīra اَلْمُغِيرَةُ.
Muḥammed مُحَمَّدٌ.
music سَمَاعٌ.
Muslim (to become a) سلم IV. — Muslim id. part. act.

Muzāḥim مُزَاحِمٌ.
naked عُرْيَانٌ fem. ة—.
name اِسْمٌ.
narrow ضَيِّقٌ (= ضَيِيقٌ) elat. أَضْيَقُ.
nature طَبْعٌ pl. § 88, 9.
near (to place) قرب II.
needle إِبْرَةٌ.
neighbour جَارٌ.
niggardly بَخِيلٌ. — to declare anyone to be n. بخل II.
niggardliness بُخْلٌ.
night لَيْلٌ.
noble كَرِيمٌ. — nobles coll. مَلَأٌ.
not see § 150.
now conj. فَ.
nutriment قُوتٌ.
O! يَا § 85; also أَيُّهَا.
obedient to (to be) قَنَتَ impf. u, with لِ.

observe (to) عبر *VIII.*
occasion (as a consequence) (to) عقب *IV, with two accus.*
Omar عُمَر.
on acount of *prep.* لِ.
one *as pronoun or adj.* وَاحِدٌ *fem.* ةٌ ـ; *with pron. suffix* أَحَدُ.
only إِنَّمَا.
onslaught بَأْسٌ.
open (to) فَتَحَ *impf. a; inf.* فَتْحٌ.
opinion (to be of) رَأَى *impf. a,* § 49 b.
or أَوْ.
other آخَرُ.
overtake (to) درك *IV.*
owner صَاحِبٌ *pl.* اِفْعَالٌ (17).
page boy غُلَامٌ.
Paradise ٱلْجَنَّةُ.

pardon (to) عَفَا *impf. u, with* عَلَى.
part (= some) بَعْضٌ (§ 133).
pass on (to let) جَازَ *med.* و *IV.*
pearls *coll.* لُؤْلُؤٌ.
pebble حَصَاةٌ.
people أَهْلٌ.
perhaps لَعَلَّ § 147 a.
perish (to) هَلَكَ *impf. i;* — to cause to p. id. *IV.*
person (man) إِنْسَانٌ.
Pharao فِرْعَوْنُ.
physician طَبِيبٌ.
pick up (to) لَقَطَ *impf. u.*
piece, see § 73 c.
piety دِيَانَةٌ.
place (occasion) مَوْطِنٌ *pl.* مَفَاعِلُ (23).
place (to) وَضَعَ *impf. a.* § 40 a.

plants coll. نَبَاتْ (masc.).
please (to) عجب IV. —
   to be well pleased with
   رَضِيَ impf. a. with عَنْ.
pleasure لَذَّة pl. § 76.
pliancy لِينْ.
poem, poetry شِعْرْ pl. أَفْعَالْ (17).
poet شَاعِرْ.
poison سمّ.
polite scholar ظَرِيفْ pl. فُعَلَاءُ (20).
poor فَقِيرْ pl. فُعَلَاءُ (20).
possessor ذُو, fem. ذَاتُ § 90 l.
poverty فَقْرْ.
power قُوَّة.
praise (to) حَمِدَ impf. a.
praise (God) ذِكْرْ.
pray (to) صلّى II.
prayer صَلَاةْ (= صَلْوَة § 43 note) pl. صَلَوَاتْ

(§ 83). direction of prayer قِبْلَة.—leader of prayer إِمَامْ.
preach to (to) خَطَبَ impf. u, with acc.
precede (to) قدم V.
prepare (to) صلح IV.
presence حَضْرَة.
present (gift) صِلَة (inf. of وصل).
preserve (to) حَفِظَ impf. a.
pride كِبَرْ.
prince أَمِيرْ pl. فُعَلَاءُ (20).
prison سِجْنْ.
promise وَعْدْ.
prophecy (gift of) نُبُوَّة. — to pretend to prophecy نبأ V.
prophet نَبِيّ pl.. أَفْعِلَاءُ (18).
prostrate oneself (to) سَجَدَ impf. u.

protect (to) حَرَسَ impf. u, i.
provide for (to) رَزَقَ impf. u, with two accus.
punishment عُقُوبَةٌ.
purify (to) طَهَّرَ II. — to oneself id V.
purse صُرَّةٌ.
pursue (to) تَبِعَ VIII.
put (to) جَعَلَ impf. a. — to put off till أَخَّرَ II. with لِ.
qualify (to) inf. وَصْفٌ.
raise, raise up (to) رَفَعَ impf. a; inf. رَفْعٌ.
Ramaḍān رَمَضَانُ.
ar-Rashīd اَلرَّشِيدُ.
reach (to) دَرَكَ IV.
read (to) قَرَأَ impf. a.
recede from (to) بَعُدَ impf. a, with مِنْ.
recite (to) قَرَأَ impf. a.
reed-pen قَلَمٌ.

reflection inf. of فَكَّرَ V.
refuse (to) أَبَى impf. a. — to r. to do, id. with أَنْ and subj.
regard, with r. to فِي.
regret نَدَمٌ.
relate (to) حَكَى impf. i.
related to قَرِيبٌ with مِنْ.
reliance inf. VIII, see rely.
religion دِينٌ.
rely on (to) وَكَلَ VIII, with عَلَى § 40 d.
repel (to) رَدَّ impf. u.
repent (to) تَابَ med. و.
repentance نَدَامَةٌ.
report خَبَرٌ.
reproach (to) لَامَ med. و.
restore (to) رَاحَ med. و IV.
restrain from (to) كَفَّ impf. u, with acc. and عَنْ. — to r. one's self from id. with عَنْ.

resurrection قِيَامَةٌ.
return to (to) رَجَعَ impf. i, with إِلَى.
right (due) حَقٌّ.
right, right hand يَمِينٌ.
rise in value (to) غَلَا impf. u.
roof سَقْفٌ.
routed (to be) هزم VII.
rule (to) سَاسَ med. و.
run races (to) سبق VIII.
saddle (of an ass) بَرْذَعَةٌ.
— saddle-girth حِزَامٌ.
safe (to be) سَلِمَ impf. a.
as-Saffāḥ اَلسَّفَّاحُ.
salt مِلْحٌ.
save (to) سلم II.
say (to) قَالَ med. و. — to say of anyone, id. with عَنْ. — to s. to anyone, id. with لِ.
scatter (to) نَثَرَ impf. u, i.

sea بَحْرٌ.
second ثَانٍ.
secret سِرٌّ pl. أَفْعَالٌ (17).
security ضَمَانٌ.
see (to) رَأَى impf. a, § 49 b.
seedy part. pass. of خمر.
seek (to) طَلَبَ impf. u. — to seek for one's self, id. V.
seemly (to be) بغى VII.
seize (to) أَخَذَ impf. u.
self نَفْسٌ § 12 e.
send (to) رسل II: for with ب; خَلْفَ.
serpent ثُعْبَانٌ.
servant (i. e. of God) عَبْدٌ pl. فِعَالٌ (9).
serve wine to (to) سَقَى impf. i, with two acc.
set about (to) صَارَ med. ى, with impf. § 99 note a.
seventy سَبْعُونَ.
shadow ظِلٌّ.

shift (to) صَرَفَ *impf. i.*
ship سَفِينَة.
shirt قَمِيص.
shoulder كَتِف.
sign آيَة *pl.* § 76.
silent (to be) *inf.* صَمْت.
sin خَطِيئَة.
singing (art of) غِنَاءٌ.
sit with (to) جلس *III, with acc.*
size قَامَة.
slave مَمْلُوك *pl.* مَفَاعِيل (27). — slave-girl جَارِيَة *pl.* فَوَاعِل (24).
sleep, go to sleep (to) نَامَ *med.* و, *impf. a; part. act. pl.* § 88, 9.
small صَغِير. — to become s. صَغِرَ *impf. a.*
smoke دُخَان.
snow ثَلْج.
so *conj.* فَ.

sober *part. act. of* صَحَا.
Socrates سُقْرَاط.
solicitude هِمَّة.
son اِبْن § 90 *b* (pluralis sanus with names of tribes).
song أُغْنِيَة *pl.* أَغَانِي (أَفَاعِل).
sorrow حُزْن.
spare (to) حَيَّ X, § 49 *c.*
speech قَوْل.
spend (to) (*of time*) *inf.* قَطْع.
spirit رُوح.
staff عَصًا.
stand (to) قَامَ *med.* و ; *part. act. pl.* § 88, 9. — to stand still وَقَفَ *impf. i.*
star نَجْم.
start off (to) وجه *f. w.* إِلَى.
stay (to) قَامَ *med.* و *IV.*
steal (to) سَرَقَ *impf. i.*
stratagem حِيلَة.

stream سَبْلٌ.
strength حَوْلٌ.
šuʻba شُعْبَةٌ.
submissive (to be) ذل V.
subsistence رِزْقٌ.
sufficiency كِفَايَةٌ.
supplication دُعَاءٌ.
surely لَ (after إِنَّ).
tail ذَنَبٌ.
take (to) أَخَذَ *impf. u.* —
  (of a city) فَتَحَ *impf. a.*
  to t. away ذَهَبَ *impf. a.*
  with بِ. — to t. hold of
  أَخَذَ *impf. u, with* بِ.
talk to (to) كلم II, *with
  acc.* — to t. to one another, id. V.
talk كَلَامٌ.
tattle أَغْطٌ.
teach (to) علم II, *with two
  accus.*

ten عَشْرٌ §§ 91, 92 a.
than مِنْ § 63 b.
that *pron.* ذٰلِكَ § 13 c.
that (in order that) لِ *with
  subj.* § 100.
that *conj.* أَنْ (*before a
  verb*) § 148 b; أَنَّ (*before
  a noun*) § 147 a.
that which مَا.
then فَ.
thereupon ثُمَّ.
thief لِصٌّ *pl.* لُصُوصٌ (10).
thing شَيْءٌ *pl.* أَفْعَالٌ (17)
  but without the nuna-
  tion أَشْيَاءُ.
think (to) ظَنَّ *impf. u, with
  two accus.; inf.* ظَنٌّ.
third ثَالِثٌ § 93 a.
thirty ثَلَاثُونَ §§ 91, 92 b.
this هٰذَا § 13 b.
Thora (the) اَلتَّوْرَاةُ.

those who مَنْ § 14 b.
thou أَنْتَ.
thought فِكْرٌ.
three ثَلَاثٌ §§ 91, 92 a.
through (by means of) prep. بِ.
throw away (to) رَمَى impf. i. — to throw down لَقَى II.
tidings, to give glad tidings to anyone of a thing بشر II, with acc. of pers. and بِ.
time زَمَانٌ.—(proper) time وَقْتٌ.
tipsy نَشْوَانُ.
title-page عُنْوَانٌ.
to (direction) prep. إِلَى; (sign of the dative) لِ.
tongue لِسَانٌ pl. أَفْعِلَةٌ (16).
towards prep. إِلَى.
transgression ذَنْبٌ pl. فُعُولٌ (10).

transitory part. act. of فَنِيَ.
travel (to) سَارَ med. ى.
treasure خَزِينَةٌ.
tree شَجَرَةٌ.
tribe قَبِيلَةٌ pl. فَعَائِلُ (25).
trick حِيلَةٌ.
trust أَمَانَةٌ.
truth صِدْقٌ.
turn (to) مَالَ med. ى. — to turn from عرض II, with عَنْ. — to t. away (act.) رَدَّ impf. u. — to t. away from (neut.) وَلَّى II, with مِنْ.
twinkling لَحْظَةٌ.
'Ubaid عُبَيْدٌ.
ugly قَبِيحٌ fem. ة.
unbeliever part. act. of كفر pl. § 76.
uncover (to) كَشَفَ impf. i.
understanding عَقْلٌ.

unmindful of (to be) سلا V, with عَنْ.

until *conj.* حَتَّى *generally with subj.* (cf. § 152 c).

upon *prep.* فَوْقَ.

used to كَانَ *med.* و, *with follg. impf.* § 99 c; *subj. gen. betw.* كان *and impf.*

value قِيمَةٌ.

vehemence حِدَّةٌ.

vehement عَاصِفٌ.

verily إِنَّ §§ 147, 96 d.

viand مَطْعَمٌ *pl.* مَفَاعِلَةٌ (28).

violent شَدِيدٌ.

wade through (to) خَاضَ *med.* و, *with acc.*

want (to) رَادَ *med.* و, *IV.*

want حَاجَةٌ.

war (holy) *inf.* فِعَالٌ *of* جهد *III.*

Waraḳa وَرَقَةٌ.

wash (to) غَسَلَ *impf. i.*

waste غَامِرٌ *fem.* ةٌ. — to render waste خرب *II.*

water مَاءٌ § 90 q.

well جُبٌّ.

well-pleased *see* please.

West مَغْرِبٌ.

whale حُوتٌ.

what *rel. interr.* مَا.

when *rel. interr.* مَتَى; *conj.* إِذَا § 158.

where? أَيْنَ. — from where, whence مِنْ أَيْنَ.

which *relat.* اَلَّذِى.

whichever أَىُّ § 14 c.

while (a) سَاعَةٌ.

whilst cf. § 157.

who *rel.* اَلَّذِى; *interr.* مَنْ.

whoever, whoso مَنْ §§ 14 b, 159.

whole جَمِيعٌ.

why? لِمَ; why then? لِمَاذَا.

wick ذُبَالَةٌ.
wickedness شَرٌّ.
will (to) شَآءَ med. ى.
wind رِيحٌ fem. § 72.
wine خَمْرٌ.
wisdom حِكْمَةٌ.
wise حَكِيمٌ pl. فُعَلَآءُ (20).
wish (to) رَادَ med. و, IV.
with مَعَ (in company w.); بِ (in union w., by means of).
without بِغَيْرِ (with gen.).
woe to! وَيْلٌ لِ.
wolf ذِئْبٌ.
woman مَرْأَةٌ, اِمْرَأَةٌ. — plur. نِسَآءٌ § 90 f.
wood عُودٌ. — piece of wood خَشَبَةٌ.

word كَلِمَةٌ.
work عَمَلٌ pl. أَفْعَالٌ (17).
world (the, this) اَلدُّنْيَا.
worst شَرٌّ § 63 note.
write to (to) كَتَبَ impf. u, with إِلَى.
wrong (to, to do) ظَلَمَ impf. i; inf. ظُلْمٌ.
Yazid يَزِيدُ.
ye أَنْتُمْ.
year سَنَةٌ pl. § 90 m.
yes نَعَمْ.
young صَغِيرٌ elat. § 63 b. pl. أَفَاعِلُ (23).
young man فَتًى pl. فِعْلَانٌ (21).
youth غُلَامٌ.
Zaid زَيْدٌ.

GLOSSARY B.

أَ *part. interr.* often before the first half of an alternative question.

أَب *st. c.* أَبُو (§ 90 *a*) father.

أَبَدَ *impf. i* to stay, remain.

أَبَدًا *adv.* always, for ever; with neg. never.

أَبَقَ *impf. i* to run away.

اِبْن *v.* بنى.

أَتَى *impf. i; c. acc.* come, come to. *c. acc. p. et* بِ *r.* to bring, to give somethg. to some one.

أَثَرَ *impf. u* to make an impression.

أَثَر *pl.* آثَار trace, sign, mark.

أَجْر wages, hire, reward.

إِحْدَى *fem.* أَحَد one, some one.

أَخ (§ 90 *c*) *pl.* إِخْوَة brother, neighbour.

أَخَذَ *impf. u* to take, to sieze, catch hold of. *VIII* to make; w. 2 Acc. to adopt, regard (as).

أَخَّرَ *II* to put off, postpone.

آخِر the last, second, end.

اَلْآخِرَة the next world.

آخَر *fem.* أُخْرَى other.

أَدَبَ *V* to conduct one's self with propriety.

أَدَب good breeding, politeness, education, polite reproof.

GLOSSARY B.

إِدَاوَةٌ vessel for holding water, made of skins.

أَدَى II to pay (tribute).

إِنْ lo! see! when lo!

إِذَا conj. when, if; adv. lo! see!

أَذِنَ impf. a; c. لِ pers. et بِ rei to allow, permit. X to ask permission.

أُذُنْ pl. آذَانْ ear.

أَذَنَ inf. I permission.

أَذَى IV to injure, molest.

اَلْأُرْدُنْ Jordan, the Jordan district.

أَرَسْطَاطَالِيسْ Aristotle.

أَرْضْ fem. earth, land, country, ground.

أَسَاسْ foundation.

أَسَرَ impf. i to tie, bind, take captive.

أَسِيرْ a captive.

اَلْإِسْكَنْدَرْ (the Arabs have treated the first two letters of the name as the article) Alexander.

اَلْإِسْكَنْدَرِيَّةُ Alexandria.

اِسْمْ v. سما.

أَصْلْ the root, the chief thing.

أُفُقْ pl. آفَاقْ region, district.

إِقْلِيمْ pl. أَقَالِيمْ (κλῖμα) region, country.

أَكَّدَ V to gather strength, become confirmed.

أَكَلَ impf. u to eat; to get to eat. III to eat with some one.

أَكْلْ inf. I eating.

مَأْكُولْ various kinds of food.

أَلَّا part. composed of أَنْ and لَا.

إِلَّا (= إِنْ لَا) except (§ 151).

اَلَّذِى fem. اَلَّتِى (§ 14a) he that; whoso, who, which.

أَلِفَ impf. a to become familiar with ...
VIII to be on intimate terms, familiarly acquainted (with).

أَلْفٌ pl. آلَافٌ or أُلُوفٌ thousand.

أَلُوفٌ intimate, familiar.

أَلِمَ impf. a to feel, suffer pain.

أَلِيمٌ painful.

إِلَهٌ pl. آلِهَةٌ a god.

إِلَهٌ et اَلٌ ex اَللّٰهُ (the true) God, Allah. عَبْدُ اللّٰهِ name of a man. اَللّٰهُمَّ O God!

إِلَى prep. (§ 96b) towards, in the direction of, to, till, as far as.

أَمْ part. interr. or.

أَمَّ impf. u, to direct one's course by something.

أُمٌّ pl. أُمَّهَاتٌ mother.

أُمَّةٌ the people of a (particular) religion, nation, people.

أَمَرَ impf. u, c. acc. p. et ب r. to order, command.

أَمْرٌ command, power; affair, matter. صَاحِبُ الْأَمْرِ commander.

أَمِيرٌ commander, prince.

أَمِيرُ الْمُؤْمِنِينَ the prince of the (true) believers, commander of the faithful = the Caliph.

أَمِنَ impf. a, c. acc. to be safe from ...
IV to believe.

أَمَةٌ pl. إِمَاءٌ female slave.

أُمَيَّةُ Umayya (man's name).

GLOSSARY B.

أَنْ (§ 100, 148b) that.

أَنْ (§§ 147, 148a) that.

إِنْ (§§ 159, 160) if.

إِنَّ (§ 147) lo! truly, verily (often untranslatable).

أَنَا pron. (§ 12) I.

أَنْتَ pron.; fem. أَنْتِ, thou.

أَنِسَ impf. a to have familiar intercourse with.

إِنْسَان coll. نَاس man.

أَنْف nose.

إِنَّمَا part. (composed of أَنَّ and مَا) only (refers in this sense usually to last word of sentence), but.

أَنَّى part. whence? how?

أَهَبَ V c. لِ rei to equip one's self, to be prepared (for any thing).

أَهْل coll. one's kinsfolk, family, people (cf.§133), inhabitants.

أَوَّل fem. أُولَى first (determ. also beginning.)

أُولُو gen. and acc. أُولِي v. ذُو.

أَيْنَ part. where? whither?

إِلَى أَيْنَ whither? مِنْ أَيْنَ (from) whence? where?

آيَة sign, revelation.

أَيُّهَا (§ 85) particle of exclamation.

بِ präp. in, on, at; with, by means of; for (of price), by (in oaths). إِذَا هُوَ بِ lo! there was...

بَابِل Babylon, Babylonia.

بَوُسَ impf. u to be brave, courageous.

بَأْس courage, strength, power.

بَحْر sea, great river.

بَدَأَ impf. a to begin.

بدل‎ *II c acc.* to exchange, alter, change.
X c. acc. et بِ to take something in exchange for (something else).

بَرِحَ‎ *impf. a* to go away, cease.

بَشَّرَ‎ *II. c acc. pers. et* بِ *r.* to tell some one something as a piece of good news.

بَصَرَ‎ *or* بَصِرَ‎ to glance, perceive; to understand something thoroughly.

بَصَرٌ‎ *pl.* أَبْصَارٌ‎ glance, intelligence.

بَطُؤَ‎ to come too late. *IV* to delay. *X* to find that sthg. comes too late.

بَطْنٌ‎ belly; bottom (of a valley).

بِطْنَةٌ‎ repletion.

بَوَاطِنُ‎ *pl.* بَاطِنٌ‎ the lowest part; the heart or secret thoughts of a person.

بَعَثَ‎ *impf. a* to arouse, awaken; to send.

بَعُدَ‎ *impf. u or* بَعِدَ‎ *impf. a* to be distant, far off. *VI* to be far distant from each other.

بَعْدَ‎ *prep.* after, after the departure, death of... مِنْ بَعْدِ‎ after the death of.

بَعْضٌ‎ one (§ 133), part, portion; some (of).

بَغِضَ‎ *impf. a* to hate.

بُغْضٌ‎ hatred.

بِغْضَةٌ‎ *id.*, state of being hated.

بَغْضَاءُ‎ hatred.

بَغَى‎ *impf. i* to seek, strive.

GLOSSARY B.

VII to be necessary, meet, behoove.

بُقْرَاط Hippocrates.

بَقِىَ *impf. a* to remain, remain over, continue in life.

بَقَآءٌ *inf.*

أَبُو بَكْرٍ Abū Bekr, name of the first Caliph.

بَكَى *impf. i* to weep.

بَلَدٌ *pl.* بِلَادٌ country, village (plur. *coll.* country).

بَلَغَ *impf. u, c. acc.* to reach, attain to; to come to one's ears.

بِلْقِيسُ Bilkīs, queen of Sheba.

بَلَا *impf. u* to try, afflict.

بَلَى *part.* certainly; nay, on the contrary.

بِمَ (*ex* بِمَا) wherewith? by what means?

بَنَى *impf. i* to build.

بِنَآءٌ *inf.*

اِبْنٌ (§ 90 *b*, بَنٍ § 6*f*. 2; 126) *pl.* أَبْنَآءٌ son.

بِنْتٌ (§ 90 *i*) daughter.

بَهِيمَةٌ *pl.* بَهَائِمُ animal, a brute beast.

بَابٌ *pl.* أَبْوَابٌ gate, door.

بَيْتٌ *pl.* بُيُوتٌ, أَبْيَاتٌ house, family. بَيْتُ ٱلْمَالِ treasury.

بَاعَ *impf. i* to sell, buy.

بَيْعٌ *inf. 1* selling, sale.

بَانَ *med.* ى *IV* to be evident.

بَيْنَ (§ 114) *prep.* between. ... بَيْنَ يَدَىْ *prop.* bet. the hands of = before, in presence of.

بَيْنَا *conj.* with a nom. sentence: while, whilst.

بَيِّنَةٌ evidence, proof.

تَابُوت masc. coffin.

تَبّ X to be well arranged, be in good order.

تَبِعَ impf. a, c. acc. to follow.
 IV c. 2 acc. to make sthg. follow, to attach sthg. to, some one.
 VIII to follow, endeavour to aquire.

تَحْتَ prep. under. مِن تَحْتِ id.

تُرَاب earth, morsel of earth.

تَرَكَ impf. u to abandon, leave, give up, omit.

تَقْوَى (cf. وقى) fem. (or. تَقْوَى msc.) piety.

تِلْكَ fem. (§ 13 c) that (woman).

تِلْمِيذ pl. تَلَامِذَة pupil, disciple.

تَمَّ impf. i to be finished.

تَمَام perfect.

تَمْرَة nom. unit. a date.

اَلتَّوْرِيَة (§ 2 d note.) the Torah (five books of Moses).

ثَأَرَ X to ask help in securing (blood) revenge.

ثَبَتَ impf. u to be or stand firm, to be fixed.
 IV to fix, establish.

ثَابِت Elat. أَثْبَتَ constant, fixed, firm.

ثَكِلَ impf. a, to lose a child (acc.) by death (said of a mother).

ثَلَاثَة fem. ثَلَاث three.

ثَلَاثَ عَشْرَة thirteen.

ثُمَّ adv. thereupon, then.

ثَنَى impf. i to bend.
 X to make an exception of.

GLOSSARY B.

ثَوْب garment.

جَاش strength of character.

جَدِيد new.

جَدْي kid.

جَذَب VIII to draw to oneself.

جَرَّ impf. u to drag, pull.

جَرَى impf. i to run, flow.

جَارِيَة pl. جَوَارٍ (§ 89) female slave, young girl.

جَزِيرَة island; اَلْجَزِيرَة Mesopotamia.

جَزَى impf. i to reward, requite.

III to pray God to requite some one for sthg.

جَسَد the body.

جَعَل to place; make, prepare; c. 2 acc. to make to be sthg.; to begin (§ 99 note a).

جَفَّ impf. i to become dry.

جَفَا impf. u to be rude.

جَفَاء inf. tyranny.

جَلَّ impf. i to be great, powerful, exalted.

جَلِيل great, illustrious, sound (in judgment).

جَلَالَة might, majesty.

جَلَس impf. i to sit down; c. لِ to give an audience.

III c. acc. to sit down by some one, sit with.

جُلُوس inf. sitting.

جَلِيس pl. جُلَسَاء companion one sits with.

جَمْرَة a live coal.

جَمَع impf. a to bring together, gather, collect.

with 2 بَيْن to bring about a meeting of two parties, to have them both come into one's presence.

أَجْمَعَ رَأْيَهُ عَلَى IV (also without رَأْيَهُ and with أَنْ) to decide upon, resolve to do sthg.
VIII to come together, to assemble.

جَمِيعٌ the whole, all (جَمِيعًا as acc. of condition: all together).

جَمَاعَةٌ a number, party (of people).

جَمُلَ to be beautiful.

جَمِيلٌ beautiful, handsome, elegant, kind.

جَنَّ impf. u to cover over, conceal.

جَنَّةٌ pl. جِنَانٌ garden of trees, Paradise.

جَنَانٌ interior, heart, soul, character.

جِنٌّ coll. demons, Jinn.

جِنِّيٌّ belonging to the demons, a demon.

جَنَبَ VIII to avoid.
فِي جَنْبِ side. جَنْبَ in comparison with.

جَنَازَةٌ pl. جَنَائِزُ corpse, funeral bier.

جَهَدَ impf. a to take trouble about sthg., exert one's self.
III to fight, do battle, esp. w. unbelievers i. e. non-Moslems.

جَهِلَ impf. a to be ignorant. جَهَالَةٌ inf.

جَاهِلٌ pl. جُهَّلٌ ignorant.

جَاهِلِيَّةٌ the state of ignorance, i. e. (pre-islamic) heathenism.

جَهَنَّمُ hell.

جَابَ med. و, IV c acc.

pers. et إلى r. to give or grant an answer, an audience to some one, listen to, promise, concede sthg. to one, comply with his request.

X to hear, in the sense of answer (a petition).

جَادَ med. و to be generous.

جَازَ med. و c. acc. to pass by.

III c. acc. to pass beyond, exceed, transgress.

جَاعَ med. و to be hungry.

جَوْعَة (nom. unit. § 73 c) hunger.

جَآءَ med. ي, c. acc. to come. c. ب to bring.

مَجِيءٌ inf.

جَيْش army.

حَبَّ IV to love.

حُبّ love.

أَحَبُّ c. (elat. حَبِيب pro dativ. pers.) pl. أَحِبَّاء dear to some one, beloved, friend.

مَحَبَّة love, friendship.

حَبَشِيّ Abyssinian.

حَبَا impf. u c. acc. pers. et ب rei to present some one with sthg.

حَتَّى until; so that; for the purpose of; (sometimes = finally).

حَجَّ impf. u to make the pilgrimage to Mecca.

حِجَج pl. حَجَّة the pilgrimage to M.

حُجَج pl. حُجَّة good reason or excuse.

حَجَبَ impf. a to prevent, exclude.

حِجَاب curtain, veil.

حَاجِب porter, gatekeeper, chamberlain.

حَدَثَ impf. u to be new. II c. acc. pers. to inform, relate. X to newly adopt, get sthg. new.

حَدِيث a story, narrative (applied esp. to the traditions respecting Muḥammed).

حَذِرَ impf. a, c. acc. vel مِن to be on one's guard against …

حَذَر inf.

حَذِقَ impf. a to be clever, skilled.

حَرَّ (حَرِرْتَ) impf. a to be free.

حُرّ pl. أَحْرَار free, noble.

حرب III to make war upon, fight with some one. VI to carry on war with each other.

حَرَدَ impf. i to strive eagerly after.

حَرَد eagerness, zeal, anger.

حرض II c. على r. to incite (to), stir up (to).

حَرَق IV to burn, singe.

حَرَّك II to move, to stir up, agitate.

حَرُمَ impf. u, c. على to be forbidden to one, to be legally prohibited one. II to pronounce unlawful, declare to be forbidden, to prohibit.

حَزِنَ to be troubled, sad. IV to trouble, make sad.

حَسَب impf. u to reckon.

حِسَاب reckoning.

حَسَد impf. u to envy.

حَسُنَ impf. u to be beautiful, good. IV to do good.

X to find to be good.

حُسْن beauty, goodness.

حَسَن elat. أَحْسَن beautiful, good.

حَشَم coll. suite, servants, escort.

حَضَر impf. u, c. acc. pers. vel عَلَى to be present with or at.

IV to bring forward, esp. to bring before a sovereign or ruler.

VIII c. acc. to come upon one (said of death). Pass. to be near to death.

حَقّ impf. u to surround.

حَفَر impf. i to dig.

VIII to dig for one's self.

حَفِظ impf. a to take care of, to guard, to be attentive.

VIII c. بِ *r.* to take care, give heed.

حَقّ impf. i to be right.

حَقّ truth, certainty; right, claim.

حَقَر impf. u to be despised. *X* to despise.

حَقِير despised.

حَكَم impf. u to decide, give judgement.

حِكْمَة wisdom.

حَكِيم pl. حُكَمَاء wise, learned.

حَاكِم pl. حُكَّام governor, ruler, judge.

حَكَى impf. i to relate.

حَلّ impf. u to loosen, untie; impf. i to be allowed.

IV or *X* to pronounce sthg. allowed, declare lawful, to allow.

حلى *V* to adorn one's self.

حَمَّ (1. pers. حَمِمْتُ) *impf. a.* to be hot.

حُمَّى *fem.* fever.

حَمَامَةٌ pigeon.

حَمَدَ *impf. i* to praise.

مُحَمَّدٌ Muhammed (the praised one).

حَمُقَ *impf. u* to be foolish.

أَحْمَقُ foolish, stupid.

حَمَلَ *impf. i* to load, carry; bring; transport. *c.* عَلَى to attack; *c. acc. pers. et* عَلَى *r.* to make s. o. sit upon sthg.; to incite to some action.

حَنِثَ to commit sin.

V to purify one's self from sin.

حَنَّطَ *II* to embalm.

اَلْأَحْنَفُ al-Aḥnaf, (a man's name).

حَاجَ *med.* و, *IV* أَحْوَجَ (§ 44 note *b*) *c.* إلَى to compel.

VIII c. إلَى to require, be in need of.

حَاجَةٌ *c.* ب need, want; *c.* إلَى request.

حَوْلَ *prep.* round, round about.

حَالٌ state, condition, situation.

حَوَى *impf. i,* to gather together, take possession (of everything).

حَىَّ *impf.* § 49 *c.* to live.

حَىُّ tribe, clan.

حَيْوَةٌ life.

خَبُثَ *impf. u* to be bad, wicked.

خَبِيثٌ bad, vile, vicious, profligate.

خَبَرَ *II c.* 2 *acc.* to relate, tell some one sthg.

VIII to test, try, prove.

خَبَرٌ pl. أَخْبَارٌ information, news, affair.

خَبِيرٌ well informed, wise.

خَبَزَ impf. i to make bread, to bake.

خُبْزٌ a cake of bread, bread.

خَتَمَ impf. i to seal up, put one's seal to.

خَدِيجَةُ Ḥadīǧa (Muḥammed's first wife).

خَدَمَ impf. u to serve.

خِدْمَةٌ inf.

خَدَمٌ coll. (the staff of) servants.

خَادِمٌ a servant.

خَرَّ impf. i to prostrate one's self, to fall down.

خَرَجَ impf. u to go out, come out, go out from, depart from.
IV to bring forth or forward, to produce, to expel.
X bring out, draw out.

خَرْجٌ tribute.

خَرِسَ impf. a to be dumb.

خَرَقَ impf. i to make a hole in, to pierce.
VII to have a hole put through, be pierced.
VIII to break through, flow through.

خَزَنَ impf. u to store up.

خِزَانَةٌ pl. خَرَائِنُ treasure, treasure-house.

خَشِيَ impf. a, c. acc. r. to fear sthg.

خَصَّ impf. u to be some one's special property.

خَاصَّةٌ, coll. خَاصٌّ an intimate friend; persons of distinction.

خَضَبَ impf. i to dye (esp. the hair).

خَضِيب dyed.

خَضِر IX to be or become green.

خَطِئ impf. a to sin.

ٱلْخَطَّاب al-Ḥaṭṭāb (a man's name).

خَفَّ impf. i to be light (opp. of heavy).

أَخَفُّ el. خَفِيف light.

خَلَدَ impf. u to be everlasting, to remain.

خَلَس VIII to appropriate to oneself secretly.

خَلَّص II c. acc. pers. et مِن to rescue, to free.

مَخْلَص escape, way of escape.

خَلَط VIII prop. to become commingled; to come on (said of the darkness in which objects can no longer be distinguished).

خَلَفَ impf. u to be behind, to succeed.
II to leave behind.

خَلِيفَة pl. خَلَفَآء Caliph.

خَلَقَ impf. u to create, form.

خَلْق 1) one's outward form; 2) coll. people.

خُلْق pl. أَخْلَاق one's (natural) disposition, character, mental and moral traits.

خَمَدَ impf. u to go out (of fire and light).

خَمِرَ to ferment.

خَمْر fem. fermented drink, wine.

خَاف med. و (§ 42 d; 44) impf. a to fear.
II to put in fear.

خَوْف fear.

خَار med. ى to be good.

GLOSSARY B.

VIII to choose, select for one's self.

خَيْر (also as *elat.*) good (adj. and noun), prosperity.

خَالَ *med.* ى II to imagine something.

دَارَا Darius.

دَبَّ *impf. i* to walk slowly.

دَابَّة *pl.* دَوَابُّ beast of burden and for riding.

دَبَرَ IV to turn one's back, go away.

دَخَلَ *c. acc.* to enter, to come; *c.* عَلَى to come to see one, to consummate marriage with (*coire*); *c.* بَيْنَ to interfere.
II to bring into, introduce.

دُخُول *inf.* I.

دَاخِل entering, future, next.

دُرّ *coll., nom. unit.* دُرَّة, pearl.

دَرَكَ IV to attain, reach, comprehend.

دِرْهَم *pl.* دَرَاهِم a dirhem, a silver coin.

دَرَى *impf. i* to know.
IV *caus.*

دَعَا *impf. u* to call, to call upon, invoke, *c.* ب to pray to God for something, to call to one's aid, to name; *c. acc. et* إِلَى to induce s. o. to do sthg., invite, summon.
VI to call to one another, *c.* ب to bring a complaint against...

دَعْوَة prayer.

دَفَعَ *impf. a* to push; hand over, deliver up.

دَنَا *impf. u, c.* مِنْ to come near.

دَنِىٌ *elat.* أَدْنَى low, humble, trivial, near; *pl.* أَدَانٍ the nearest parts.

دُنْيَا *fem.* world.

دَاخَ *med.* و II to subdue.

دَارَ *med.* و *c.* لِ to surround.

دَارٌ *pl.* دُورٌ dwelling-place, house, abode, court.

دَامَ *med.* و to remain, continue, be durable.

دُونَ *prep.* on this side of, below, beneath; other than, exclusively of, besides, before. مِنْ دُونِ id.

دَوِىَ *impf. a* to be indisposed. IV to treat medically.

دَوَاءٌ medicine.

دَانَ *med.* ى to be in subjection.

دَيْنٌ debt.

دِينٌ *pl.* أَدْيَانٌ religion.

دِينَارٌ denar, a gold coin.

ذَا *pron.* (§ 13 a) this. مَاذَا (§ 15) what (then)?

ذِئْبٌ wolf.

ذَعَرَ *impf. a, c. acc.* to frighten.

ذَكَرَ *impf. u, c. acc.* to think of, mention, name, speak of. Inf. ذِكْرٌ.

ذَلَّ *impf. i* to be insignificant, feeble.

ذَلِيلٌ miserable, feeble.

ذَلِكَ *fem.* تِلْكَ *pron.* (§ 13 c) that.

ذَهَبَ *impf. a* to go, go away. IV to cause to disappear.

ذَهَبٌ gold.

ذُو the (man) of, possessor of cf. §§ 90 l, 133.

ذَاعَ med. ى to become known, spread abroad. IV to make public, publish.

رَأْس pl. رُؤُوس head, the chief thing.

رَئِيس pl. رُؤَسَاء leader, general.

رَأَى impf. يَرَى (§ 49 b) to see, be of opinion, think, believe, consider advisable, c. 2 acc. to regard or esteem a person or thing as, hold to be. IV أَرَى c. 2 acc. to show.

رَأْى insight, counsel, advice.

رَبّ lord, God.

رَبَطَ impf. u to tie, fasten. رَبِيط elat. أَرْبَط securely fastened, firm.

رَبْع pl. رِبَاع house, pl. real estate.

اَلرَّبِيع ar-Rabī', (a man's name).

أَرْبَع fem. أَرْبَعَة four.

رَجَعَ impf. i to turn back, return.

رُجُوع inf.

رِجْل fem. pl. أَرْجُل foot, leg.

رَجُل pl. رِجَال a man.

رَجَمَ impf. u to stone.

رَجِيم stoned, accursed.

رَجَا impf. u, c. acc. to hope for sthg.

رَحُبَ to be wide, broad. II c. ب to bid anyone welcome (مَرْحَبًا).

رَحِمَ impf. a, c. acc. pers. to have pity on, compassion for, some one.

I to take compassion on each other.

رَحْمَة loving kindness (esp. of God), deed of kindness.

رَحًى *fem.* mill.

رَخُوَ *vel* رَخِىَ to be flaccid, soft.

رَدَّ *impf. u* to bring back, give back.
VIII to turn back.

رَدّ *inf. I* giving back.

رَزَقَ *impf. u, c. 2 acc.* to present, grant, furnish, bless with, give food.

رِزْق food (esp. as given by Allah), sustenence.

مَرْزُوق Marzûḳ, (man's name).

رسل *IV* to send.

رَسُول *pl.* رُسُل messenger, apostle (esp. of God).

رَصَّعَ *II* to set (of jewels), inlay.

رَضَعَ *impf. a* to suck (at the breast).
IV to give suck.

رَضِىَ *impf. a, c. acc.* to be content with, acquiesce in, take pleasure in.

رَضِىَ ٱللّٰهُ عَنْهُ God be gracious unto him!
IV to satisfy, render content.

رِضًى *inf. I* pleasure, delight (in sthg.).

رَعَدَ *VIII* to shake, tremble.

رَعَى *impf. a* to watch, tend.

رَاعٍ *pl.* رُعَاة herdsman, shepherd.

رَعِيَّة *pl.* رَعَايَا subjects (also *sing. coll.*).

مَرْعَاة *pl.* مَرَاعٍ pasture-ground.

رَغِبَ *impf. a* to have a strong craving for; *c.* عَنْ to give up the craving for sthg., to shun, relinquish.

رَغِيفٌ *pl.* أَرْغِفَةٌ (flat) cake.

رَفَعَ *impf. a* to raise, lift up (the voice); *c.* إِلَى to bring sthg. before the judge.

رَفِيعٌ high, noble.

رَفَقَ *IV c.* بِ to be kind, gentle with . . .

مِرْفَقٌ *pl.* مَرَافِقُ elbow.

رَقَّ *impf. i* to be or become thin, abject, mean.

رِقٌّ bondage, slavery.

رَقَعَ *impf. a* to mend, patch.

رُقْعَةٌ patch.

رَكِبَ *impf. a, c. acc.* to mount on horse-, camel-back &c., to ride.

رُكُوبٌ *inf.* stepping into, aboard (a ship).

رَمَضَانُ name of a month.

رَمَى *impf. i, c.* بِ *r.* to throw, pelt with.

رَهِبَ *impf. a, c. acc. rei* to be afraid of sthg.

رَاهِبٌ monk.

رَاحَ *IV med.* و, *c. acc. el* مِنْ to rid . . . of.

رِيحٌ (for رُوحٌ) *fem.*, *pl.* رِيَاحٌ wind.

رَائِحَةٌ smell, scent.

رَادَ *med.* و, *IV c. acc.* to will, wish, intend, endeavour to.

رَامَ *med.* و to seek, desire, attack.

رَوَى *impf. i* to relate.

زَحَفَ *impf. a* to advance slowly.

زَرَعَ *impf. a* to sow.

زَرْع *coll.* seed, green corn, green crop, different sorts of grain.

زَعْزَعَ to shake violently. *II* (reflexive).

زَعَم *impf. u* to assert, relate.

زَفّ *impf. u* to conduct a bride to her husband's house.

زَكَا *ult.* و to increase, to be good, pure.

أَزْكَى *elat.* زَكِىّ pure, delicate, dainty.

زَلْزَلَ to shake (trans.). *II* to shake (int.), tremble.

زَمّ *impf. u* to fasten securely.

زِمَام bridle (nose-rein).

زَمَان time, space of time.

زَهَا *impf. u* to shoot up, to flourish, prosper.

زَهْو *inf.*

زَاج *med.* و *II c.* 2 *acc. vel c. acc. et* مِن *vel* لِ to marry some one to, join in wedlock; *c. acc.* to take in marriage. *V c. acc.* reflex.

زَار *med.* و *impf. u* to visit. زِيَارَة *inf.*

زَال *med.* و *impf. a* to cease. زَوَال *inf.* cessation. Noon *or* afternoon.

زَوَى *impf. i* to remove, clear away.

زَاوِيَة *pl.* زَوَايَا corner.

زَاد *med.* ى *impf. i, c.* 2 *acc.* to give more, to add to. مَزِيد *inf.;* increase, addition.

سَ *part.* § 95 *d*; 99 *a*.

سَيَّر *impf. a* to be or remain over.

Glossary B.

سَادِر remaining, the rest, all.

سَأَل impf. a, c. 2 acc. to ask one for sthg. c. acc. pers. et عَن to enquire for, ask respecting.

سَائِل beggar.

مَسْأَلَة the asking, a question.

سَبِيل masc. or fem. way, right way, road.

سِت fem. سِتَّة six.

سَتَر impf. u or i to hide, shield (e. g. from the gossip of the people).

سَجَد impf. u, c. لِ, to prostrate one's self προσκυνεῖν.

سُجُود inf.

مَسْجِد mosque.

سَرّ IV c. إِلَى pers. to tell s. o. sthg. as a secret.

سِرّ pl. أَسْرَار secret.

سَرْج pl. سُرُوج saddle.

سَرُعَ IV to be in haste, c. فِى to make haste with...

سَرِيع, elat. أَسْرَع, quick, swift, speedy.

سُرَاقَة Surāka, (a man's name).

سَطَح impf. a to spread out.

سَطْح the flat roof of eastern houses.

سَاعِد pl. سَوَاعِد the forearm.

سَعَى impf. i, c. بِ vel فِى pers. to lodge information against, denounce.

سَفَر pl. أَسْفَار journey.

سَفِينَة ship.

سَكَت impf. u to become or be silent.

سَكِرَ impf. a to be or become drunk.

سَكَارَى pl. سَكْرَانُ drunk.

سَكَنَ impf. u to dwell, inhabit, rest, be quiescent.

سُكُون rest, quiescence.

سَاكِن pl. سُكَّان inhabitant.

سَلْسَلَ to put in chains.

سَلُطَ impf. u to be or become powerful. II to make, install as ruler.

سُلْطَان c. عَلَى authority over, rule; ruler, sultan.

سَلِمَ impf. a to be whole, intact. II to bestow health and prosperity; c. على to greet, salute. IV c. لِلَّهِ to declare one's self resigned to to God; to become a Moslem.

سُلَّم ladder.

سَلَام immunity from ills, prosperity, welfare. عَلَيْهِ ٱلسَّلَامُ peace be with him! (parenthetically placed after the names of high religious personalities).

سَلَامَة peace and prosperity.

إِسْلَام (inf. IV) Islām.

سَمَّ impf. u to put poison into anything, to poison.

سَمّ poison.

سَمِعَ impf. a to hear.

أَسْمَكَة pl. سَمَك fish.

سَمَّى II c. 2 acc. vel c. acc et بِ to call by name to give a name to.

اِسْم (§ 56 a) name.

سَمَاء heaven.

GLOSSARY B.

سَنَّ *impf. u* 1) to sharpen, 2) ordain, institute.

سِنٌّ tooth, age.

سُنَّةٌ *pl.* سُنَنٌ regulation, institution, tradition (of the Moslems).

سَنَدَ *IV* to support.

سَنَةٌ *pl. nom.* سِنُونَ (§ 76 b; 90 m) year.

سَهَرَ *impf. a* to keep awake.

سَهَرٌ *inf.*

سَاءَ *med.* و to be bad, wicked. *IV* to spoil, corrupt, to do ill.

سَاخَ *med.* و to sink into the ground.

سَادَ *med.* و *c. acc.* to become lord, ruler, over...

أَسْوَدُ *fem.* سَوْدَاءُ *pl.* سُودٌ, سُودَانٌ black.

سَيِّدٌ *pl.* سَادَةٌ lord, ruler, chief.

سَاعَةٌ hour, short space of time, moment.

سَوْفَ *part.* § 95 d; 99 a.

سَاقَ *med.* و to drive.

سُوقٌ *pl.* أَسْوَاقٌ market, bazaar, lane.

سَوِيَ *VIII* to be equal, alike, simultaneous with.

سَوَاءٌ *c.* عَلَى (quite) the same, indifferent to.

سَارَ *med.* ى *impf. i* to journey, go along, go. فِي أَثَرِ to follow one's track.

مَسِيرَةٌ distance travelled.

سَيْفٌ *pl.* سُيُوفٌ, أَسْيَافٌ sword, sabre.

شَاءَمَ *VI* to find a bad omen.

شُؤْمٌ a bad omen.

شَبِعَ *impf. a* to be satiated. *IV* to satiate, satisfy.

شَبَّهَ *II* to compare.

شَجَرَ *impf. u* to be intricate, intertwined.

شَجَرَةٌ *nom. unit.* tree, shrub.

شَدَّ *impf. u* to bind, tie.
II c. عَلَى to press hard on one.
VIII to become strong, powerful, heavy.
شَدِيدٌ *elat.* أَشَدُّ strong, powerful; vehement.

شَرَّ (1. pers. شَرِرْتُ) *impf. a* to become bad.
شَرٌّ (*elat. id.*) pl. أَشْرَارٌ bad, wicked. Mischief, woe, war.

شَرِبَ *impf. a* to drink.
شَرَابٌ wine, strong drink.

شَرُفَ *impf. u* to be high.
IV to be high, lofty.
شَرَفٌ height, fame, nobility.

شَرِيفٌ pl. أَشْرَافٌ noble, aristocratic, respected.

شَرَقَ *impf. u* to rise (of the sun).
مَشْرِقٌ place of the sun's rising = the East.

شَرِكَ *impf. a c. acc.* to be one's companion.
شَرَكٌ net.
شَرِيكٌ companion, ally.

شَرَى *impf. i* to buy, sell.
VIII to buy, negotiate.

شَغِفَ *impf. a, c.* بِ *pers.* be deeply struck with.
شَفَةٌ pl. شِفَاهٌ lip.
شَفِقَ *IV c.* عَلَى to be tenderly solicitous for...

شَكَرَ *impf. u* to thank, be thankful.

شَكَا *impf. u*, to complain.
VIII to complain.

شمت *II c. acc.* to say "God

bless you" to a person (e. g. sneezing).

شَمْس fem. sun.

شِمَال left (hand or side).

شَهِدَ impf. a c. acc. to be present at sthg., to witness, to give evidence. III c. acc. to see, be an eye-witness. شَاهِد pl. شُهُود witness.

شَهَادَة testimony, guarantee, security, a bearing testimony.

شَهْر pl. أَشْهُر month.

شها VIII to desire, wish.

شَهْوَة sensual desire, appetite.

شَاة coll. nom. unit. شَاة small cattle, sheep and goats; nom. unit. a single head of these.

شَار med. و, IV c. إِلَى to point to.

شَوَى impf. i to roast.

شَاء med. ى impf. a to will, wish.

— شَيْء a matter, thing, something.

شَابَ med. ى to become gray-haired.

شَيْب gray hairs.

شَاخَ med. ى to become an old man.

شَيْخ old man.

شَيْطَان pl. شَيَاطِين devil, Satan.

شَاع med. ى to spread abroad, become public. II to accompany, to follow.

صَبّ impf. u to pour, pour out.

صَبُحَ impf. u to be attractive, good-looking. IV to enter the time of

Socin, Arabic Grammar²   I

early morning. *c. acc.* become sthg. early, soon.

صَبَاحَةٌ beauty, loveliness.

صَبَرَ *impf. i, c.* عَلَى to have patience with, to put up with, endure.

صَبَغَ *impf. u or a* to dye.

صَبَا *impf. u* to be foolish; youthful.

صَبِيٌّ *pl.* صِبْيَانٌ little boy.

صَحَّ *impf. i* to be in good health, sound.

أَصَمُّ *pl.* صِحَاحٌ *elat.* صَحِيحٌ right, correct.

صَحِبَ *impf. a, c. acc.* to keep company with, have to do with.

III to take for companion.

X to take with one as an associate.

صَاحِبٌ *pl.* أَصْحَابٌ associate; friend, companion; owner, inhabitant of (cf. § 133).

صَحِيفَةٌ *pl.* صُحُفٌ leaf.

صَدَّ *impf. u* to turn away from, alienate.

صَدَقَ *impf. u* to speak the truth, be truthful, sincere.

II to consider sthg. to be true, right, to believe one.

V c. عَلَى *pers. et* بِ *rei* to give one sthg. as alms.

صَدِيقٌ *pl.* أَصْدِقَاءُ friend.

صَرَفَ *impf. i* to turn from.

VII to turn, go, away, return (home).

صَاعِقَةٌ thunderbolt.

صَغُرَ *impf. u,* to be small, little.

صِغَر *inf.* littleness.

صَفَّ *impf. u* to place in a row, draw up.

VIII to arrange (themselves), to stand in a row.

صفر *IX* to be yellow.

أَصْفَر *pl.* صُفْر yellow.

صَفَى *VIII* to choose.

مُصْطَفَى man's name.

صَلَحَ *impf. a* to be good, be in order.

IV to put in order, set right.

صَالِحَة a pious action, good deed.

صلّى *II* to pray, perform divine service, to worship. صَلَّى ٱللّٰهُ عَلَيْهِ وَسَلَّمَ contracted to صَلْعَمَ § 11.

صَلْوَة, صَلَاة divine service, worship, prayer.

صَمَتَ *impf. u* to be quiet.

صُمُوت silence.

صَنَعَ *impf. a* to make, prepare, to do.

صِهْر *pl.* أَصْهَار relation (by marriage).

صَابَ *med.* و, *IV* to befall, fall to one's share.

مُصِيبَة misfortune.

صَوْت voice.

صُورَة *pl.* صُوَر figure, shape, form.

صَوْمَعَة cell.

صَاحَ *med.* ى to cry out.

VI to shout at each other.

صَادَ *med.* ى to hunt.

صَيْد *inf.* hunting, what is caught, game.

صَارَ *med.* ى, *c. acc.* to become or be sthg.; to repair to.

II to cause to become; to appoint, to place.

ضَجَعَ VIII to lie on one's side.

ضُحًى forenoon.

ضَرَبَ impf. i to strike, beat.

    VIII refl. to beat against each other.

ضَرْب inf. I striking, beating.

ضَرْبَة a single blow, a beating.

ضَعُفَ impf. u to be weak.

ضَعِيف weak.

ضَلَّ impf. i to err.

ضَلَالَة erring, error.

ضَمَّ impf. u to put close to, press against, to gather.

ضَآءَ med. و to be clear, bright, shining.

ضَوْء light, brightness.

ضِيَاء brightness.

ضَيْف pl. أَضْيَاف guest.

ضَاقَ med. ى IV to press hard, hem in.

ضِيق straits, distress.

طَأْطَأَ to sink (trans.).

طَبَّ impf. u or i to treat medically.

طَبِيب physician, doctor.

طَبَق IV to cover with a lid.

    VII to be covered up.

طَحَنَ impf. a to grind.

طَحِين flour.

طَرَدَ impf. u to chase away, drive away, pursue.

طَرَقَ IV to cast down one's eyes.

طَعِمَ impf. a to eat.

    IV to feed (trans.).

طَعَام inf. I eating, taste, a meal, food, a (particular) dish.

طَلَبَ *impf. u* to seek, search after; wish for.

طَلَبْ *inf. I* seeking, a search.

أَبُو طَالِبٍ Abū Ṭālib (Muḥammed's uncle).

طَلَعَ *impf. u* to stand up, get up, rise (of the sun).

VIII c. عَلَى to look at, see.

طَلَّقَ *II c. acc.* to set free, give divorce to.

IV to set free.

VII to go away, depart.

طَمِعَ *impf. a* to strive to obtain, to covet, sthg.

طَمَعْ *inf.* covetousness, greed.

طَهُرَ *impf. u* to be clean, pure.

II to cleanse, purify.

طَاعَ *med.* و to obey, be compliant.

IV id.

طَاعَةْ *inf.*, obedience, subjection.

طَافَ *med.* و to go round.

طُوفَانْ flood.

طَالَ *med.* و *IV* to lengthen, protract; to be long over sthg.

طَوِيلْ long, lasting long.

طَوَى *impf. i* to fold, fold up *or* together.

طَابَ *med.* ى to be good, pleasant, excellent.

أَطْيَبْ *Elat.* طَيِّبْ good, excellent, nice to the taste, sweet (scent).

طَيِّبَة something good, a dainty.

طَارَ *med.* ى to fly.

طَانَ *med.* ى *II* to plaster with clay or mud (طِينٌ).

ظَلَمَ *impf. i* to treat unfairly, injure, do wrong to.

IV to grow dark.

ظُلْمَةٌ darkness.

ظَلَامٌ darkness, dusk.

ظَلَّامٌ one that acts injuriously, oppressor.

ظَهَرَ *impf. a* to appear, to come in sight.

IV to bring to sight.

ظَهْرٌ the back, upper part, surface.

ظَاهِرٌ *elat.* أَظْهَرُ prominent, striking.

عَبَدَ *impf. u* to worship.

عَبْدٌ *coll.* عَبِيدٌ slave, servant; *pl.* عِبَادٌ man (as the servant of God).

عَبْدُ ٱللّٰهِ 'Abdallāh (a man's name).

عِبَادَةٌ adoration, worship.

عَبَرَ *impf. u* to cross, to pass along (a certain road).

عِبْرَةٌ an example (from which to take warning).

عَبَسَ *impf. i* to look stern, black-browed.

اَلْعَبَّاسُ al-'Abbās, (man's name).

عَبَاءَةٌ mantle, cloak.

عَتَقَ *impf. i* to be or become free.

IV to free, liberate.

عِتْقٌ nobility, high rank.

عَتِيقٌ free, noble, old.

عُثْمَانُ 'Utmān, (man's name).

عَجِبَ *V c.* مِنْ to wonder at sthg.

GLOSSARY B.

عَجِيبَة pl. عَجَائِب a wonder, miracle.

عَجَّل II to expedite.

عَدَّ impf. u, c. 2 acc. to count, reckon as ...

عَدَس lentils.

عَدَل impf. i to be just.

عَادِل just, impartial.

عَدِم impf. a, c. acc. to be without sthg.

عَدَا V c. acc. to cross over, go beyond.

III c. acc. to treat as an enemy, attack.

عَدُوّ pl. أَعْدَاء enemy.

عَدَاوَة enmity.

عَذَّب II to torture, punish.

عَذَاب torture, punishment.

عَذَر VIII to excuse one's self.

عُذْر excuse (in the sense of a refusal).

عَرَب coll. the Arabs.

أَعْرَابِيّ a Bedouin.

عَرَض impf. i to interfere with, thwart, offer.

V to come in one's way.

عَرَف impf. i to perceive, know, recognize.

مَعْرِفَة knowledge.

مَعْرُوف a favour, kind deed.

اَلْعِرَاق name of the country known to the ancients as Babylonia.

عَزَّ impf. i to be strong, powerful (often parenthetically after اَللّٰه: he is powerful).

أَعَزّ elat. عَزِيز strong, powerful.

عَزَل impf. i to depose.

VIII to take one's leave, be deposed.

عَزَى c. acc. et عَنْ rei II to console, comfort.

عَسْكَرُ pl. عَسَاكِرُ a body of troops, army.

عَشَّ II to build a nest (عُشّ).

عَاشَرَ III to associate with.

عَشْرُ fem. عَشَرَةُ ten.

عَشِيرَةُ pl. عَشَائِرُ tribe, tribesmen.

مَعْشَرُ assembly, the whole; those present.

عَصَى impf. i, c. acc. to resist, not obey some one.

مَعْصِيَةُ pl. مَعَاصٍ resistance, revolt, sin.

عُضْوُ member.

عَطَسَ impf. i or u to sneeze.

أَعْطَى IV c. acc. pers. et rei to give sthg. to some one.

عَطِيَّةُ pl. عَطَايَا gift, present.

عَظُمَ impf. u to be or become great, large; c. عَلَى to appear to be great, insolent.

عِظَمُ inf. greatness.

عَظِيمُ elat. أَعْظَمُ great, of great account, august.

عِفْرِيتُ pl. عَفَارِيتُ a wicked, clever demon.

عَفَا impf. u, c. عَنْ to pardon (a person), be gracious to.

عَقْرَبُ pl. عَقَارِبُ scorpion, a bitter enemy.

عَقَلَ impf. i. or عَقِلَ impf. a to be intelligent.

عَقْلُ intellectual ability, intelligence; prudence.

عَلَّ VIII to fall ill.

عِلَّة illness, sickness.

عَلَّجَ III to treat.

عَلِمَ impf. *a* perceive, know, learn (that). *c.* بِ to know something. *IV c. 2 acc.* to acquaint, inform one of sthg.

عِلْم *pl.* عُلُوم knowledge, science.

عَلَامَة mark, sign.

عَالِم *elat.* أَعْلَم *pl.* عُلَمَاء possessing knowledge, a learned man, *savant*.

عَلَّام very knowing.

مُعَلِّم teacher.

عَلَا *impf. u* to be high. *VI* to be highly exalted, esp. parenthetically after *Allah*: He is exalted (§ 23).

عَلَى *prep.* (§ 96 *b*) over, on the ground of, on, upon, at; with verbs of entering: *chez*; against, in the direction of, towards. هُوَ عَلَى شَيْ to be in a state of, to be accustomed to sthg.

عَلِي *elat.* أَعْلَى high; also man's name 'Ali.

عَالٍ *elat.* أَعْلَى high, prominent, excellent.

عَمَّ *impf. u* to be or become common; to increase.

عَمّ uncle (on the father's side); اِبْنُ ٱلْعَمّ cousin.

عَامَّة the common people (plebs), large crowd.

عَمَّرَ *II* to furnish, provide handsomely.

عُمْر life; in the oath لَعَمْرِي by my life.

عُمَر 'Omar (man's name).

عَمْرُو (§ 90n pronounce 'Amrun) 'Amr (a man's name).

عَمِلَ impf. a to do, make, construct.

X to employ one for for some purpose, to apoint governor.

عَمَلٌ pl. أَعْمَالٌ work, act, deeds of piety, province.

عَامِلٌ pl. عُمَّالٌ a functionary, vicegerent, prefect.

عَمِىَ impf. a to be or become blind.

IV to disfigure, make unrecognisable.

أَعْمَى pl. عُمْىٌ blind.

عَنْ prep. away from, from (hinderance); about, concerning; according to, on the authority of.

عِنَبٌ pl. أَعْنَابٌ vine, grape.

عِنْدَ prep. by the side of, near, with, by (one).

عَاجَ med. و IX to be bent, crooked.

عَادَ med. و to return, c. acc. to visit.

عَاذَ med. و c. ب to take refuge in...

X to ask for protection; to say: أَعُوذُ بِٱللّٰهِ (Surah 114) "I take refuge in God", c. مِنْ from.

عَانَ med. و IV c. acc. to help, support.

X to help one's self, to help on, succour.

مُعَاوِيَةُ Mu'āwiya, the first Omayyad Caliph (661—679).

عِيسَى Jesus.

عَاش‎ *med.* ى‎ to live.
عِيشَة‎ life, way of living, (§ 64 c).
غَدَا‎ *ult.* و‎ to come early. V to breakfast, to refresh oneself early.
غَرَب‎ *impf. u* to set (of the sun).
مَغْرِب‎ place where the sun sets, the West.
غَرِق‎ IV to make to sink, drown.
غَزَال‎ gazelle.
غَسَل‎ *impf. i* to wash.
غَشِى‎ *impf. a* to cover.
غَاشِيَة‎ *pl.* غَوَاشٍ‎ saddle-cover, horse-cloth.
غَصَب‎ *impf. i c. acc. rei et* مِنْ‎ *p.* to take sthg. from one unlawfully.
غَضِب‎ *impf. a* to get angry, be angry with.

غَفَر‎ *impf. i c.* لِ‎ *pers.* to pardon, forgive.
مَغْفِرَة‎ pardon, forgiveness.
غَفَل‎ *impf. u* to neglect.
غَفْلَة‎ inattention, negligence.
غَلَب‎ *impf. i* to be all-powerful, victorious.
غلق‎ II *et* IV to bolt, bar, shut.
غُلَام‎ *pl.* غِلْمَان‎ a young man, lad, slave.
غَنِى‎ *impf. a, c.* عَنْ‎ to be rich.
غَنِىّ‎ *pl.* أَغْنِيَآء‎ rich.
غنى‎ II to sing.
غَار‎ *med.* و‎ to penetrate far into, go down.
غَار‎ a cave.
غَاص‎ *med.* و‎ to dive.

غابَ *med.* ى to be absent, c. عَنْ to disappear. غَيْبٌ *pl.* غُيُوبٌ a secret.

غَيْبَةٌ absence, stay among strangers.

مَغِيبٌ *inf.* sunset.

غارَ *med.* ى II to alter, change.

غَيْرٌ (§ 133 with gen.) another, somethg. different from, no (with neg.), except; before substs., adjs. and parts. it renders the converse, like our prefix *un-* or *in-*; مِنْ غَيْرِ without.

فَ *conj.* (§§ 95 e; 152; 161) and so, then, and.

فَتَحَ *impf. a* to open. *VII* to open (intr.). *VIII* to conquer, acquire for one's self.

فَتْحٌ *inf.* I.

مِفْتَاحٌ *pl.* مَفَاتِيحُ key.

فَتَنَ *VIII c.* بِ to be struck with emotion, bewitched, by.

فَتًى a young man.

فَتَاةٌ a young woman, girl.

فَجَرَ *impf. u* to transgress, act viciously.

فَاجِرٌ *pl.* فُجَّارٌ evil-doer.

فَخَرَ *impf. a* to boast of, glory in. *III* to give oneself airs towards some one.

فَخْرٌ *inf.* I.

فَرَّ *impf. i* to flee.

اَلْفُرْسُ the Persians.

فَارِسُ Persia.

فَرَسٌ a horse, esp. of a good breed.

فَرَشَ *impf. u* to spread out.

فِرَاش pl. فُرُش carpet, cushion, bed.

فَرَض impf. i, c. على pers. to impose sthg. on one as a duty.

فَرَغ imp. u, c. مِن to be empty, disengaged, finished with sthg.

فَرَق impf. u, to separate, part.

III to leave.

VIII to become separated, to disperse.

فَزِع impf. a to get a fright, be afraid.

فَسَد impf. u to become bad, wicked.

فَسَاد inf. the doing of mischief, evil, wrong.

فَشَا IV to divulge, publish, betray.

فَضَل impf. u to be or remain over, to be excellent.

فَضْل bounty, kindness, favour.

فَطِن impf. a to be clever.

فِطْنَة intelligence.

فَعَل impf. a to do.

فِعْل pl. أَفْعَال deed, act, mode of action.

فَقَد V to miss, enquire for, some one.

فَقِير pl. فُقَرَآء poor.

فَكّ VII to free oneself, to become disattached.

فكر V to reflect.

فَاكِهَة pl. فَوَاكِه fruit.

فلح IV to become happy, successful, to prosper.

فُلْك, فُلُك a (large) ship.

فُلَان so and so, Mr. Such-and-Such.

فَلَاة pl. فَلَوَات desert.

فَم mouth (§ 90 o).

فَاتَ med. و c. acc. to pass

by, to expire (of the time for some one to do sthg.).

فَاقَ *med.* وَ *to excel, be excellent.*

فَوْقَ *prep.* above, higher than.

فُوهٌ (§ 90o *pl.* أَفْوَاهٌ) mouth.

فِي *prep.* in, into, at, on, among, accompanied by, by; with (before a quality), in relation to, with regard to.

فَيْلَسُوفٌ *pl.* فَلَاسِفَةٌ philosopher.

قَبُحَ X to find detestable.

قَبْرٌ *pl.* قُبُورٌ grave.

قَبَضَ *impf. i* to take hold of, take into one's hand.

قَبِلَ *impf. a* to accept.

IV to approach, come nearer; be susceptible to.

V to receive.

X to be opposite.

مِنْ قَبْلُ *adv. vel* قَبْلُ before.

قَبْلَ *prep.* before.

قِبَلَ *prep.* in the presence of, in the sphere of ... مِنْ قِبَلِهِ on his side, of his party.

قَبُولٌ *inf. I* acceptance.

قَبِيلَةٌ tribe, family (in wide sense).

مُقَابَلَةٌ comparison, relation.

قَتَلَ *impf. u* to kill, make away with.

III c. acc., to fight with, fight.

قَتْلٌ *inf. I* killing, execution.

قَتِيلٌ *pl.* قَتْلَى killed.

أَبُو قُحَافَةَ Abū Ḳuḥāfa, the father of Abū Bekr.

قَدْ (§ 98 e, 99 d) *particle*.

قَدَرَ *impf. i* to be able to, can, could (also with folg. impf.). *c.* عَلَى to have power over.

IV c. عَلَى to make one more powerful than...

قَدْرٌ worth, value, due, power. بِقَدْرِ in relation to, in proportion to ...

قَدِمَ *impf. a*, to advance, approach.

II to place before, set sthg. before s. o.

IV to approach.

V to go before, precede.

قَدِيمٌ *pl.* قُدَمَاءُ ancient, old, of a past time.

قَرَّ *impf. i* to stay, persevere.

IV to render stable, *c.* بِ *rei* to confess to sthg.

X to stand fast, hold good.

قَرَارٌ continuance, rest.

قَرَأَ *impf. a* to read.

قُرْآن Ḳur'ān *or* a passage therefrom.

قَرُبَ *impf. u* to be near at hand.

II to place near, to take as intimate friend, to offer, set before one.

VI to be close together.

قَرِيبٌ *pl.* أَقْرِبَاءُ; *elat.*

أَقَارِبُ *pl.* (subst.). *c.* مِنْ near, close (to); related.

قُرَيْشٌ the tribe of the Ḳuraish, the Ḳuraishites.

قُرَشِيٌّ *nom. rel.* a Ḳuraishite.

قَرْنٌ horn; ذُو ٱلْقَرْنَيْنِ the two horned (Alexander bicornis).

قَرْيَة pl. قُرًى place, village.

قَسَا ult. و to be hard.

قَسَمَ IV to swear.

قَصَدَ impf. i, to make for, repair to, some one.

مَقْصِد the end of a journey.

قَصُرَ to be short.

VI to shorten one's self, to shrink.

قَصْر pl. قُصُور castle, fortress.

قَضَّ VII to let one's self down, dart down (of a bird).

قَضَى impf. i to decide judicially; to accomplish, finish; to discharge a claim.

VII to be finished, brought to an end.

قَضَاء inf. I payment.

قَطُّ adv. ever, with negat. never.

قَطَعَ impf. a to cut off.

VII c. عَنْ to become parted from; to cease.

قَعَدَ impf. u to seat one's self, sit down.

قُفْل pl. أَقْفَال lock, padlock.

قَلَّ impf. i to be small, few.

IV to make small, take little of.

X to deem small, think little of, despise.

قَلِيل small, few, scant.

قَلَبَ impf. i to turn round, to change.

VII to alter (intr.), to change one's mind.

قَلْب pl. قُلُوب heart.

قَلَعَ VIII to tear away, take away.

قَنَصَ impf. i to hunt, catch.

قَنَا VIII to procure, purchase.

قَادَ *med.* و to lead, guide. *VII* to let one's self be guided.

قَالَ *med.* و to say, tell; often = ask. *c.* ل to name.

قَوْلٌ *pl.* أَقْوَالٌ speech, utterance, apothegm.

مَقَالٌ speech.

قَامَ *med.* و to stand up, proceed (to). *IV* to fix, set up, establish; halt, stop, stay. *X* to be upright, faithful.

قَوْمٌ *coll.* people, one's dependants, nation, subjects.

قِيَامَةٌ resurrection.

قَائِمَةٌ *pl.* قَوَائِمُ foot.

مَقَامٌ place, occasion.

قَوِيَ *impf. a* to be strong.

قُوَّةٌ strength, force;

c. بِ the means to do sthg.

قَوِيٌّ (*c.* عَلَى) strong, powerful.

كَ (§§95*f*; 145*b* prop. subj.) as, like as.

كَأَنَّ (it is) as if . . .

كَبَرَ *impf. u* to be great, large. *V* to vaunt oneself, be proud.

كِبَرٌ *inf. I* to be advanced in years.

أَكْبَرُ *elat.* كَبِيرٌ great, old.

كَتَبَ *impf. u* to write. *III c. acc.* to correspond with.

كِتَابٌ *pl.* كُتُبٌ a writing, scripture (= written revelation), letter, book.

كَتَمَ *impf. u* to conceal.

K

كِتْمَان *inf.* concealing, keeping close.

كَثُرَ to be much *or* many.

IV to make many, take much of.

X to consider much *or* many.

كَثِير *elat.* أَكْثَرُ much, many (often rather as a subst. in apposition).

كَذَبَ *impf. i* to lie, tell lies.

II c. acc. pers. vel بِ *rei* to charge one with falsehood, discredit.

كِذْب *inf. I,* lying, a lie, falsehood.

كَرَبَ *impf. u,* to cause one trouble, pain.

كُرْبَة grief, distress, anxiety.

كُرْدُوس *pl.* كَرَادِيس division (of cavalry), squadron.

كَرُمَ *impf. u* to be noble, generous.

كَرِيم *pl.* كُرَمَاء noble, high-souled, highly esteemed.

مَكْرُمَة *pl.* مَكَارِم a noble quality, generous action.

كَرِهَ *impf. a* to dislike.

كَسَبَ *VIII* to acquire, to attain to sthg.

كَسَفَ *impf. i* to eclipse.

كَشَفَ *impf. i, c.* عَنْ to uncover.

VII to be uncovered, be carried off.

كَعْب ankle-bone, a die (*pl.* dice).

كَافَأَ *III c. acc. pers. et* عَلَى *rei* to requite, recompense one for sthg.

كَفَرَ *impf. u* to be unthankful, to deny.

كَافِر pl. كُفَّار unbelieving.

كَفَنَ impf. u to wrap in a shroud.

كَفَى impf. i, c. acc. pers. et r. to do sthg. in some one's place; to protect s. o. from sthg.

كُلّ (§ 119 b) totality; before determ. subst., all; before indeterm., every.

كُلَّمَا as often as ...

كلم II c. acc. pers. to speak with, address one. V to speak, talk, make speeches. c. ب to pronounce, utter.

كَلِمَة word.

كَلَام speech, talk, conversation.

كَمْ (§ 15) how much?

كَمَا (كَ + مَا) as.

كَمِيل perfect.

كَمِين impf. a to hide one's self.

كَنْز pl. كُنُوز treasure.

كَنَا impf. u to give one a surname contg. أَب.

كَانَ med. و to be, exist. (Sometimes the perf. of this verb is to be translated by our present). c. acc. (§§ 110, 149) to be something. c. لِ to be translated by "to have".

مَكَان pl. أَمْكِنَة place.

كَيْفَ how?

لَ (§§ 95 g; 147 b) a corroborative particle.

لِ prep. (§§ 95 h; 117; 130; 131; 132) for; is sign of the dative; on account of, for ... sake (giving purpose, motive); at (the time of).

لِ conj. c. subj. (§ 100)

K*

in order that; *c. mod. apoc.* § 101 a. لِأَنْ (§ 147) because.

لَا (§§ 101 b; 111; 150 c) not, no. بِلَا *prep. c. gen.* without. By means of a preceding negation لَا is very frequently resumed.

(وَلٰكِنْ) لٰكِنْ, لَاكِنْ (often) nevertheless, but.

مَلَاكٌ (also مَلَكٌ) *pl.* مَلَائِكَةٌ angel.

لَبِثَ *impf. a* to tarry, delay.

لَبِسَ *impf. a* to put on. IV *c.* 2 *acc.* to clothe.

مَلْبَسٌ, *pl.* مَلَابِسُ clothing, dress.

لَحِقَ *impf. a c.* بِ or *c. acc.* to overtake.

لَذَّ, 1. *pers.* لَذِذْتُ, *impf. a* to be tasty, sweet.

لَذِيذٌ *elat.* أَلَذُّ tasty, delicious, sweet.

لَزِمَ *impf. a, c. acc.* to remain in ...

لِسَانٌ *pl.* أَلْسُنٌ tongue.

لَصِقَ VIII to cling to.

لَطُفَ *impf. u* to be fine, slender, kind.

لُطْفٌ *c.* بِ kindness, graciousness, towards...

لَطِيفٌ kind.

لَعِبَ *impf. a* to play, sport.

لَعَلَّ (§ 147) may be, perhaps.

لَعَنَ *impf. a* to curse.

لَعْنَةٌ a curse.

لَقَّبَ II *c. acc. pers. et* بِ to surname, give a nickname to.

لَقِمَ *impf. a* to swallow, gulp down.

لُقْمَةٌ a morsel.

لَقِىَ *impf. a* to meet, meet with.
  IV c. acc. to throw. X to throw one's self, to lie.

لَمْ (§ 101 c) not.

لَمَّا *conj.* after, when.

لَوْ *part.* if, introduces a condition, which is not likely to be fulfilled.

لَامَ *med.* و to blame.

لَوْن *pl.* أَلْوَان colour, sort, kind.

لَيْسَ (§§ 50; 110; 144) not to be, to be non-existent.

لَيْل, لَيْلَة *pl.* لَيَالٍ (§ 90 p) night.

مَا, مَا ذَا *pron.*(§ 15) what? (§ 14) that which, what, somethg. that.

مَا *conj.* (§ 158 b) so long as.

مَا not (cf. § 150).

مَأْرِب Ma'rib, a town in South Arabia.

مِثْل resemblance, likeness; the like, same; one (pers. or thing) like, cf. § 145 b.

مَثَل resemblance, nature, quality (of a thing).

مَحَنَ *impf. a* to put to the test.

مِحْنَة *inf.*

مَدِينَة *pl.* مُدُن, مَدَائِن town. اَلْمَدِينَة = مَدِينَةُ النَّبِيِّ Medina.

مَرَّ *impf. u c.* بِ to pass by.

مَرَّة "time". مَرَّة once. *pl.* مِرَارًا often.

مَرَارَة bitterness.

اِمْرُؤ (§ 90 e) man.

اِمْرَأَة woman, wife.

مُرُوءَة manliness, virtues, manly virtue.

مَرَازِبَة pl. مَرْزُبَان margrave, prefect.

مَرِض impf. a to be or become sick.

مَرِيض sick.

مرغ V to roll (in the dust).

مَرْيَم Miriam, Mary.

مَسَح impf. a c. بِ to wipe, wipe off, away.

ٱلْمَسِيح Christ, the Messiah.

مَسَك I to take hold of, sieze.

V to hold on by sthg.

مَسَا IV to enter on the eventide; to do something late.

مَشَى impf. i to go, walk;

مَشْى inf.

مَضَى impf. i to go, betake one's self to.

مُضِى (§ 71 e) inf I.

مَطَر pl. أَمْطَار rain, shower of rain.

مَطَل impf. u to defer (a payment).

مَطْل inf.

مَع prep. with; besides; alongside of.

مَعِدَة stomach.

مَقَت impf. u to hate.

مَقْت hatred.

مَكَّة Mecca.

مَكَث impf. u to tarry, stay.

مَلَأ impf. a, c. acc. et مِن to fill sthg. with ...

VIII to become filled.

مَلَك impf. i, c. acc. to rule, govern, possess.

II to appoint as king.

مُلْك dominion, sovereignty, reign, riches.

مِلْك possessions, riches.

مَلِك pl. مُلُوك king.

مِلَاك ceremony of marriage.

مَمْلَكَة pl. مَمَالِك kingdom, sovereignty.

مَلَكَ for مَلْأَك v. under لأك.

مِمَّن = مِن + مَن (§ 5 note b).

مَن who? (§ 15); he who, they that; one that, whoso, whoever (§§ 14, 154, 159).

مِن prep. of (= some of, in partitive sense § 114), belonging to; with the negation it has a strengthening effect, § 141; consisting of; away from, from (separation, point of departure); hence in comparison = than; through (passage).

مُنْذُ (مِن ذُو from) since.

مَنَع impf. a, c. 2 acc. to debar one from sthg., refuse, prevent one doing sthg.; c. acc. et مِن to defend one from or against sthg.

VIII to protect one's self.

مَهَر impf. a, c. بِ to be skilled, clever, expert, wellversed.

مَهْر wedding-present, price of the bride (paid to her father).

مَات med. و to die.

II to put to death.

مَوْت inf. death.

مَيِّت dead.

مُوسَى Moses.

مَال pl. أَمْوَال goods and chattels, property, flocks.

مَآءٌ (§ 90 q) pl. مِيَاهٌ water.

مَائِدَةٌ pl. مَوَائِدُ table, tray.

مَازَ med. ى II c. بَيْنَ to distinguish.

نَبَأَ II c. acc. pers. et ب rei to give one information regarding.
V to give one's self out for a prophet.

نَبَهَ VIII to awake up.

نَبِىٌّ pl. أَنْبِيَآءُ vel نَبِيُّونَ prophet.

نُبُوَّةٌ the office, rank, of prophet.

نَجُدَ impf. u to be brave, courageous.

نَجْدَةٌ courage, magnanimity.

نَجْمٌ pl. نُجُومٌ constellation.

نَجَا impf. u to become free, to save one's self.

IV causative.

نَحْنُ pron. we.

نَحَا V to turn aside, to draw back, retire.

نَخْلٌ coll., nom. unit. نَخْلَةٌ palm.

نَدِمَ impf. a, c. عَلَى to repent of sthg., feel sorry.

نَدِمَ III to be one's boon companion.

نَدِيمٌ pl. نُدَمَآءُ boon companion, mess-mate.

نَدَا III call out, c. acc. to call to some one.

نَذَرَ IV to warn.

نَزَعَ impf. i to remove.
VIII to strip off, displace.

نَزَلَ impf. i to descend, alight, stop, lodge, encamp. c. عَلَى to alight at, lodge, stay with...
IV to send down (in

particular, a revelation).

مَنْزِلٌ pl. مَنَازِلُ dwelling-place, abode, halting-place.

نَسَخَ impf. a to copy.

نُسَخٌ pl. نُسْخَةٌ a copy.

نَسِىَ impf. a to forget.

نِسْيَانٌ inf. forgetting.

نِسَآءٌ (§ 90 f) women.

نشد III c. 2 acc. to adjure by God.

نَشِطَ impf. a to be lively, in good spirits.

نَشَاطٌ inf.

نَصَبَ impf. u to set up.

نَصِيبٌ share, portion.

نَصَحَ impf. a to be a true friend.

نَصَرَ impf. u, c. acc. to help, succour.

V (denom.) to become a Christian, to live as a Christian.

نَصَارَى pl. نَصْرَانِىٌّ a Christian.

اَلْمَنْصُورُ al-Manṣūr, the second Abbaside Caliph 754—775.

نَضَا VIII to draw (the sword).

نَطَحَ impf. a to butt with the horns.

نَطَقَ impf. i to talk.

IV to make, compel to talk.

نَظَرَ impf. u to see, look at, examine, reflect.

نَعِمَ impf. a to be soft, well off, affluent.

نَعَمٌ coll. a herd of camels.

نِعْمَةٌ affluence, welfare.

نَعَمْ part. yes, yes indeed.

نَفَرَ *impf. u vel i c.* مِنْ to flee from, avoid.

نفس *II* to cheer, relieve. نَفْسٌ *fem., pl.* أَنْفُسٌ, نُفُوسٌ soul (anima appetens), self (§ 12e); life. قَتْلُ نَفْسٍ بِغَيْرِ نَفْسٍ the taking of a life not for a life, i. e. without a murder having been committed.

نَفَعَ *impf. a* to be of use. *VIII c.* بِ make use of, profit by . . .

مَنْفَعَةٌ *pl.* مَنَافِعُ use, useful qualities, benefit.

نَفَقَ *III* to play the hypocrite.

نَقَمَ *impf. i, c.* مِنْ *pers.* to reproach one with sthg. *VIII* to avenge one's self.

نَقِمَةٌ an act of revenge.

نَكَبَ *impf. u* to afflict, hurt, injure.

نَكْبَةٌ affliction, trouble.

نَكَحَ *impf. i* to marry. *III id.*
X id., to wish to marry. نِكَاحُ ٱلْمَقْتِ marriage with one's stepmother.

نكد *V* to be hard, strait, troublesome.

نَكَرَ *IV* to deny. *c. acc. r. et* على to find strange, to take offence at sthg.

نَهُدَ *impf. u* to be fat, large.

نَهْدٌ large, aspiring, generous.

نَهْرٌ *pl.* أَنْهَارٌ stream.

نَهَى *impf. a* to forbid. *VIII* to arrive at, come to an end.

نُوحٌ Noah.

نَار *fem. pl.* نِيرَان fire, hell-fire.

نُور light.

نَوْع *pl.* أَنْوَاع kind, species, different (sort of).

نَوْفَل a man's name.

نَاقَة *pl.* نُوق female camel.

نَام *med.* و, *impf. a* to lie down, sleep.

هَجَر *impf. u* to part from some one.

اَلْهِجْرَة, هِجْرَة the removal of Muhammed from Mecca to Medina.

اَلْهَدْهَان Hadhād, name of a king.

هَدَى *impf. i* to lead by the right way, to guide aright.

اَلْمَهْدِيّ al-Mahdī, name of the third Abbaside Caliph, 775—785.

هٰذَا, *fem.* هٰذِهِ (§ 13 b), this, here.

هَرَب *impf. u* to flee.

هَزَم *impf. i* to put to flight. *VII* to turn and flee.

هَاشِم Hāšim, man's name; بَنُو هَاشِم Muḥammed's clan.

هَلْ *part. interrog.*

هُمْ, هُمُ *pron. 3. pers. plur. msc.* they (§ 12 a).

هَمَّ *impf. u* to intend to do sthg.

هِمَّة energy.

اَلْهِنْد India, the Hindus.

هُوَ *pron.* he.

هَار *med.* و, *VII* to collapse.

هَان *med.* و, to be easy. *X c.* ب to despise.

هَوَان insignificance.

هَوِى *impf. a, c. acc.* to fall in love with.

هَوَآءٌ air, sky.

هِيَ pron. III fem. she.

وَ conj. and, also, even. Asseverative particle w. the genit.: وَٱللّٰهِ by God (be it sworn). c. acc. with (§ 112).

وَثَنٌ pl. أَوْثَانٌ an idol.

وَجَبَ impf. i to be necessary; to be legally incumbent on one.

IV to necessitate.

مُوجِبٌ vel مُوجِبَةٌ (part. act. IV) pl. مَوَاجِبُ that which brings about sthg., occasion, cause.

وَجَدَ impf. i to find.

وجه V to take the direction of . . ., set out.

وَجْهٌ pl. وُجُوهٌ face, countenance.

وَاحِدٌ one, single.

وَحَى IV c. إِلَى pers. to reveal to one, inspire.

وَدَّ impf. a to love.

VI to love mutually.

مَوَدَّةٌ love, inclination.

وَدَعَ impf. يَدَعُ to set, place, leave, let.

II to deposit.

IV c. acc. rei et إِلَى pers. to intrust sthg. to some one.

وَدِيعَةٌ pl. وَدَائِعُ property given in trust, a deposit (of money or its equivalent).

وَرِثَ impf. يَرِثُ to inherit.

VI to receive as one's portion.

وَارِثٌ heir.

وَرَدَ impf. i to go down, arrive.

وَرَقَةُ Waraḳa, man's name.

وَزِير pl. وُزَرَآءُ vizier, minister.

وَسِخَ impf. a to be dirty.

وَسَخ inf.

وَسِعَ impf. يَسَعُ to be possible, be open (to one). IV to bring one into a comfortable position; to get riches for s. o.

وَسِنَ impf. a to be sleepy.

وَصَفَ impf. i to describe.

صِفَة description.

وَصَلَ impf. i to connect, arrive at. VI to be mutually attached to each other.

وَصَّى IV to bequeath by will.

وَصِيّ executor (of a will).

وَضَعَ impf. يَضَعُ to lay. VI c. لِ to be humble, to appear humbly before...
VIII to be humbled, powerless.

وَضِيع low, ignoble, mean.

مَوْضِع pl. مَوَاضِع place, position, dwelling-place.

وَعَدَ impf. i to make an agreement, promise. VIII to accept a promise, to promise one another.

مِيعَاد rendezvous, appointed time.

وَعَظَ impf. i to warn, exhort. VIII to suffer oneself to be corrected.

وِعَآء pl. أَوْعِيَة vessel, receptacle.

وَقَدَ impf. i to go forth

to a prince. *c.* على to come to.

وَفَقَ *III c. acc.* to agree with, correspond to.

وَفَى *impf. i* to be complete.

III c. acc. to come to, arrive at.

تَوَفَّاهُ ٱللّٰهُ, تَوَفَّى *V* God has taken him (the Moslem) to himself, has brought him to a blessed end. *Pass.* to die a blessed death.

وَفَاةٌ dying; a blessed end.

وَقْتٌ time.

وَقَعَ *impf.* يَقَعُ to fall, fall upon, light upon; *c.* عَلَى to find some one.

IV to excite.

وَقَفَ *impf. i* to stop, stand; *c.* عَلَى to go up to one.

وَقَى *V c. acc.* to beware, be afraid, of sthg.

VIII to be afraid.

وَكَّلَ *II* to appoint as overseer. *V* to trust (in).

وَكِيلٌ representative, vice-gerent, agent.

وَلَدَ *impf. i* to bring forth. *IV c. acc.* to beget. *X c. acc.* to beget (a son) by a woman.

وَلَدٌ *pl.* أَوْلَادٌ child, son, lad. (In the sing. also *collect.*).

وَلِيمَةٌ feast, marriage feast.

وَلِيَ *impf. i, c. acc.* to be near.

II to turn one's back, to turn round; *c.* عَنْ to turn away from.

وَلِيٌّ *pl.* أَوْلِيَاءُ near; esp. 'near to God' = saint, helper.

مَوْلًى *pl.* مَوَالِي client, slave.

وَهَبَ اِمpf. يَهَبُ c. 2 acc. to present some one with sthg.).

يَا part. of exclam. (§ 85) O!

يَبِسَ impf. a, c. مِنْ to despair of...

يَتِيمٌ pl. أَيْتَامٌ orphan.

يَثْرِبُ Yaṯrib, name of Medina before Islam.

يَدٌ fem., pl. أَيْدٍ (§ 90 r) hand, power, possession.

يَسَرَ impf. i to play (either with arrows, by wh. lots were cast, or with dice).

مَيْسِرٌ play, game, game of chance.

يَقِظَ impf. a to be awake. IV to wake. X to have one's self waked, to awake.

يَمِينٌ on the right, the right side, right hand.

اَلْيَهُودُ coll. the Jews.

يُوسُفُ Joseph.

يَوْمٌ pl. أَيَّامٌ (§ 90 s) day, pl. length of reign. يَوْمَ on the day that...(§ 129). يَوْمًا one day; with suff. e. g. يَوْمَكَ thy day (§ 125). اَلْيَوْمَ (§ 118 a) to-day. يَوْمَئِذٍ (= يَوْمَ إِنْ) in that day, then.

يُونَانِيٌّ a Greek.

CORRIGENDA.

pp. 56, 57 for headings as printed read: § 65 Nomina Relativa; § 66 Nomina Deminutiva.

p. 68 heading read: § 78 Nom. Diptota.

p. 93 l. 4, for 'you' read 'them'.

p. 40* 4, read ومُنْذِرِينَ.

p. 42*, 7 read بَاخٍ.

p. 42*, 15 read صِرْتَ.

p. 54*, 2 read أَرِسْطَاطَالِيسُ.

p. 55*, 7 read أَمَرْتَهُمْ.

REUTHER & REICHARD, PUBLISHERS, BERLIN.

PORTA LINGUARUM ORIENTALIUM

sive

ELEMENTA LINGUARUM

Hebraicae, Phoeniciae, Biblico-Aramaicae, Samaritanae, Targumicae, Syriacae, Arabicae, Aethiopicae, Assyriacae, Aegyptiacae, Copticae, Armeniacae, Persicae, Turcicae, aliarum

studiis academicis accommodata ediderunt

J. H. Petermann, H. L. Strack, E. Nestle, A. Socin, F. Prätorius, Ad. Merx, Aug. Müller, Frdr. Delitzsch, C. Salemann, V. Shukovski, Th. Nöldeke, Ad. Erman, G. Steindorff, R. Brünnow, Dav. Heinr. Müller, K. Marti, H. Zimmern, G. Jacob, alii.

"**Porta Linguarum Orientalium**" is a Library of Oriental Grammars edited by the most eminent German Scholars, adapted for self-study and text-books in Colleges.

Their most commendable characteristic is that they offer to the beginner in one book:

1. A concise but complete Grammar, as far as possible modelled on the same plan for all Languages. For the Semitic languages differences and conformities are specially pointed out. All volumes published since 1885 contain a chapter on Syntax also.

2. A Chrestomathy. For the purposes of comparative philology and to avoid an otherwise necessary interlinear version in seven grammars the first four chapters of Genesis according to the old translations form the initial chapter. In some volumes reading exercises precede the Chrestomathy proper.

3. A Dictionary, explanatory of the words found in the Grammar and Chrestomathy.

4. A Bibliography, carefully selected, suggestive of advanced work.

Where necessary special and separately published Chrestomathies are included in the plan of the "Porta."

The following volumes are now ready in English, French, German or Latin Editions:

LONDON: WILLIAMS & NORGATE, 14, Henrietta Street.
NEW YORK: B. WESTERMANN & Co., 812, Broadway.

REUTHER & REICHARD, PUBLISHERS, BERLIN.

Arabic: Arabic Grammar, Paradigms, Literature, Chrestomathy and Glossary by *A. Socin*. 2nd edition. 1895. Cloth. 8 6.
Arabische Grammatik, mit Paradigmen, Litteratur, Chrestomathie und Glossa: von *A. Socin*. (Vol. IVa). Dritte Auflage. 1894. Paper. 6/— net.
Arabic (Prose) Chrestomathy, with Glossary by *R. Brünnow*. 1894. Cloth. 8 6
Delectus veterum carminum arabicorum, carmina selegit et edidit *Th. Noeldeke* Glossarium confecit *A. Mueller*. 1890. Cloth. 7 6.
Arabic Bible-Chrestomathy, with a Glossary, edited by *George Jacob*. 1888 Cloth. 2 6.

Ethiopic: Aethiopische Grammatik, mit Paradigmen, Litteratur, Chrestomathie und Glossar von *Franz Prätorius*. 1886. Paper. 6 — net.
Grammatica aethiopica cum Paradigm., Litteratura, Chrestomathia et Glossario scripsit *Franz Prätorius*. 1886. Paper. 6,— net.

Hebrew: Hebräische Grammatik, mit Übungsbuch von *H. L. Strack*. Fünfte durchges. Auflage. 1893. Cloth. 4/— net.
Hebrew Grammar with Reading book, Exercises, Literat. and Vocabulary by *H. L. Strack*, second enl. ed. 1889. Cloth. 4.6.
Grammaire hébraique avec Paradigmes, Exercices de lecture, Chrestomathie e Bibliogr. par *H. L. Strack*. 1886. Paper. 3/6 net.

Samaritan: Grammatica samaritana, Litteratura, Chrestomathia cum Glossari edidit *J. H. Petermann*. Editio secunda emendata. Paper. 4/ net.

Aramaic (Chaldee): Grammatica chaldaica, Litteratura, Chrestomathia cur Glossario edidit *J. H. Petermann*. Editio secunda emendata. Paper. 4,— ne
Grammatik des Biblisch-Aramaischen mit Chrestomathie von *Karl Mart*. (In preparation).

Targumic: Chrestomathia targumica edidit adnotat. critica et glossario instrux *Ad. Merx*. 1888. Paper. 7/6 net.

Syriac: Syriac Grammar with Bibliography, Chrestomathy and Glossary b *Eb. Nestle*. 1889. Cloth. 9/—.
Syrische Grammatik mit Litteratur, Chrestomathie und Glossar von *Eb. Nestle* Zweite verm. und verbess. Auflage. 1888. Paper. 7,— net.

Assyrian: Assyrian Grammar with Paradigms, Chrestomathy, Glossary an Literature by *Friedr. Delitzsch*. 1889. Cloth. 15/—.
Assyrische Grammatik mit Paradigmen, Übungsstücken. Glossar und Litteratu von *Friedr. Delitzsch*. 1889. Paper. 12,— net.

Egyptian: Egyptian Grammar with Paradigms, Chrestomathy, Glossary an Literature. 1894. Cloth. 18 —.
Altaegyptische Grammatik mit Litteratur, Chrestomathie und Glossar vo *Ed. Erman*. 1894. Paper. 16/— net.

Coptic: Koptische Grammatik mit Litteratur, Chrestomathie und Glossar vc *G. Steindorff*. 1894. Cloth. 14,/— net.

Armenian: Grammatica armeniaca, Litteratura, Chrestomathia cum Glossar. edidit. *J. H. Petermann*. Paper. 4 — net.

Persian: Persische Grammatik mit Paradigmen, Litteratur, Chrestomathie ur Glossar von *C. Salemann* und *V. Shukovski*. 1889. Paper. 7/ net.

Turkish: Türkische Grammatik mit Paradigmen, Litteratur, Chrestomathie ur Glossar von *August Müller*. 1889. Paper. 8,— net.

Supplement: Lehrbuch der Neuhebräischen Sprache und Litteratur von *H. L. Strac* und *C. Siegfried*. Paper. 3 — net.

LONDON: WILLIAMS & NORGATE, 14, Henrietta Street.
NEW YORK: B. WESTERMANN & Co., 812, Broadway.

REUTHER & REICHARD, PUBLISHERS, BERLIN.

SÎBAWAIHI'S
BUCH ÜBER DIE GRAMMATIK
nach der Ausgabe von H. Dérenbourg
und dem Commentar des Sîrâfî

übersetzt und erklärt
und
mit Auszügen aus Sîrâfî und anderen Commentaren versehen
von
Dr. G. Jahn,
Professor in Königsberg.

Mit Unterstützung der Königl. Preuss. Akademie der Wissenschaften und der Deutschen Morgenländischen Gesellschaft.

In zwei Bänden. Lex.-8⁰.

Bis jetzt erschienen: I. Band 1. u. 2. Hälfte, Uebersetzung § 1—145
Erklärungen § 1—145.

Subscriptions-Preis: Mark 32.—

ORIENTALISCHE BIBLIOGRAPHIE
begründet von August Müller.
Unter Mitwirkung von
Th. Gleiniger-Berlin, G. Grotenfelt-Helsingsfors, G. Kalemkiar-Wien,
J. V. Prášek-Kolín, C. Salemann-Petersburg, H. L. Strack-Berlin,
K. V. Zetterstéen-Upsala

bearbeitet	herausgegeben
von	von
Dr. Lucian Scherman,	**Dr. Ernst Kuhn,**
Privatdoc. an der Universität in München,	Professor in München.

Mit Unterstützung der Deutschen Morgenländischen Gesellschaft.

Preis pro Jahrgang (Band) M. 10.—.

Erschienen sind bis jetzt Band I—VIII. (1887—1894).

„Man kann fest behaupten, dass etwas Ähnliches an umfassender und erschöpfender Sorgfalt auf dem Gebiete der oriental. Literatur noch niemals geboten worden ist."
Prof. C. SIEGFRIED (Jena) im Theol. Jahresbericht VIII

LONDON: WILLIAMS & NORGATE, 14, Henrietta Street.
NEW YORK: B. WESTERMANN & Co., 812, Broadway.

REUTHER & REICHARD, PUBLISHERS, BERLIN.

BIBLIOTHECA GEOGRAPHICA PALAESTINAE.

Chronologisches Verzeichniss

der

auf die Geographie des heiligen Landes bezüglich. Litteratur

von 333 bis 1878

und

Versuch einer Cartographie.

Herausgegeben von

Reinhold Röhricht.

Mit Unterstützung der Gesellschaft für Erdkunde zu Berlin.

gr. 8. XX. 744 S. M. 24.—.

Ein Werk von geradezu einzigartiger Bedeutung. Die gesammte wissenschaftliche Kritik (*Berliner Zeitschr. für Erdkunde — Studien aus d. Benediktiner-Orden — Theol. Lit. Blatt — Theol. Lit. Zeit. — Theol. Jahresbericht — Theol. prakt. Quart. Schrift — Zeitschr. d. Deutsch. Pal. Ver. — Schweizer Blätter f. Wissenschaft u. Kunst — Lit. Rundschau f. d. Kath. Deutschl. — Lit. Centralbl. — Oest. Lit. Centralbl. — Görres Hist. Jahrb. — Centralbl. f. Bibliothekwesen — Götting. Gel.-Anzeigen — Lit. Handweiser — Wiener Zeitschr. f. Kunde d. Morgenl. — Sunday School Times — Palestine Explor. Fund — Jewish Review — Athenaeum — Church Quart-Review — Scottish Geogr. Magazine — Proceedings of the Royal Geogr. Soc. — Revue Bénédictine — Archives Israélites — Revue d. quest. histor. — Revue histor. — Revue critique — Biblioth. de l'école de chartes — Université cathol. — Archivio storico italiano — Journ. d. Russ. Minist. für Volksaufklärung — Hist. Quart.*) ist darin übereinstimmen, dass die Vollständigkeit sowie die Treue und Gewissenhaftigkeit der ganzen Arbeit eine durchgängig vollendete ist und dass ein ähnliches Werk von gleichhervorragendem wissenschaftlichen Werth nicht existirt. Es dürfte für alle, welche sich mit der Erforschung Palästina's beschäftigen, in welcher Richtung es auch sein möge, ein geradezu unentbehrliches Nachschlagebuch bilden, jede grössere Bibliothek müsste dasselbe besitzen.

LONDON: WILLIAMS & NORGATE, 14, Henrietta Street.
NEW YORK: B. WESTERMANN & Co., 812, Broadway.